Veda - The Supreme Science of Creation

Robert E. Wilkinson

Acknowledgements:

Cover Illustration: The Inner Chamber: Photographs by Kai Sievertsten of William Netter's model of The Inner Chamber of the Mother's original Temple plan.

Published by Æon Books
P. O. Box 396
Accord, NY 12404
http://www.aeongroup.com
ISBN-13: 978-0945747017

Dedication

To Thea, Light of the World.

Skambha Intensive 2005, at the Falls
Aeon Centre of Cosmology, Kodaikanal, India

Table of Contents:

Introduction

'The Veda was the beginning of our spiritual knowledge; the Veda will remain its end. These compositions of an unknown antiquity are as the many breasts of eternal Mother of knowledge from which our succeeding ages have all been fed. The recovery of the perfect truth of the Veda is therefore not merely a desideratum for our modern intellectual curiosity, but a practical necessity for the future of the human race. For I believe firmly that the secret concealed in the Veda, when entirely discovered, will be found to formulate perfectly that knowledge and practice of a divine life to which the march of humanity, after long wanderings in the satisfaction of the intellect and senses, must inevitably return.' [1] **Sri Aurobindo, (*India's Rebirth*, p. 94)**

The Vedas are the world's oldest and most profound scriptures. This collection of ancient mantras, prayers and hymns emerged out of the Indian mind and spirit during a period estimated by historians to be between 1700 and 1100 BCE. Some knowledgeable authorities, however, believe that the Vedas are much older and have their origin in immense antiquity. The texts are organized into four cosmic divisions known as Samhitās. These include the Rigveda, the Yajurveda, the Samaveda and the Atharvaveda. Its name *'Veda'* comes from the Sanskrit root *'vid'* which means 'to know' or 'to see'. This is not the kind of knowledge obtained through our ordinary conceptual understanding. The Knowledge contained in the Vedas is of a much higher order known as *'Śruti'*, or *'that which is heard'*. Śruti is a direct revelation of truth in the soul, a preexistent inner knowledge as opposed to knowledge acquired through empirical observation and experience. Owing to their sublime origins, the Vedas have been described by the poet-seer Sri Aurobindo as *'...the highest spiritual truth of which the human mind is capable'*[2]. For thousands of years these truths lay fallow and unexplored because most Sanskrit scholars believed that the prayers and hymns were the superstitious utterings of primitives and barbarians concerned with only the most material aspects of life. It was not until the 20th century and the discoveries of Sri Aurobindo, the Mother and Thea, [Patrizia Norelli-Bachelet] that its complex metaphysics were finally unveiled. What one is astonished to discover is that the Vedas represent a unique view into the nature of existence itself. Its Cosmological ideas reveal an advanced knowledge of the laws of the universe that far exceeds our modern scientific understanding. Like the Pyramids of Giza, the Vedas are one of the greatest mysteries of the ancient world because they reveal a depth of knowledge that a 'primitive' civilization, according to our present view of history, could not possibly have known.

While the Vedas contain a rich pantheon of gods and goddesses, they do not celebrate a central spiritual figure or prophet like Jesus, Buddha or Muhammad. The Vedic deities represent the powers, personalities and archetypes of the Supreme Godhead. The Rishis were primarily concerned with the realization of these godheads in the individual as an aid and an inspiration on the journey to *Swar* - the Vedic Truth-Consciousness.

> *'...In these four verses of the opening hymn of the Veda we get the first indications of the principal ideas of the Vedic Rishis - the conception of a Truth-Consciousness supramental and divine, the invocation of the gods as powers of the Truth to raise man out of the falsehoods of the mortal mind, the attainment in and by this Truth of an immortal state of perfect good and felicity and the inner sacrifice and offering of what one has and is by the mortal to the Immortal as the means of*

> *the divine consummation. All the rest of Vedic thought in its spiritual aspects is grouped around these central conceptions.'*[3] **Sri Aurobindo, [CWSA] *The Secret of the Veda, P. 64***

Along with the help of the Gods, the knowledge of Evolution in Time is an indispensable aid for those who would plumb the deepest mysteries of the Veda. Its secret wisdom, obtained through practice of an ancient yoga, reveals that the initiatic journey of the Rishis is a journey in Time through the Year and its Twelve Months (the central figure of the Veda around which the Sacrifice is conducted). Part and parcel of this Vedic realization is a recognition of the profound connections and correspondences that exist between the Cosmic order and the patterns of its manifestation on the Earth.

> *'...The key to the difference between the Veda and what followed, is the point Sri Aurobindo makes when he states that dristi, sruti, smriti, ketu formed the foundation of the Vedic experience. These are illuminations that arise solely within one's inner being, never through dissolution of the nexus of one's consciousness by extension to the Beyond. Yet though this Earth-orientation is the primary ingredient in a true Vedic quest, its very first premise, it is entirely overlooked today.'*[4]**Thea [Patrizia Norelli-Bachelet] 'Sri Aurobindo and the Condition of Vedic Wisdom in India', February, 2007 [underlined emphasis mine]**

In the pages that follow we will examine the state of our earthly evolution and the multiple crises we face today that can trace their origins to the 'otherworldly' dispositions imposed upon our mental species by the partial and incomplete religions and philosophies of the past. Starting with Buddhism, every spiritual realization in the last two thousand years has lured us farther away from the sacred Earth with promises of heavens, nirvanas and paradisiacal afterlives. In a world where personal 'salvation' is defined in terms of escape, negation or illusion, what, may we ask, is the purpose of the Earth? Is it simply a meaningless stepping stone to another, truer reality? Or, is the Earth like us, a living evolving being moving toward a more meaningful union with the Divine? The conclusion that I hope will become irrefutably clear is that the only way through and out of these evolutionary crises and on to a more enlightened existence is a return to the eternal wisdom of the Vedas, because they, more than any other spiritual path, lead the seeker toward an Earth centered and Earthly realized spirituality.

VEDA - The Supreme Science

In the early years of the last century India's greatest sage, Sri Aurobindo, published a series of essays that were later compiled into his principle work of spiritual philosophy entitled, *The Life Divine*. The overarching theme of his 1130 page opus is that Man is a transitional being. He is not final. In man and high beyond him ascend the radiant degrees that climb to a Divine Supermanhood. The next approaching achievement in the Earth evolution, according to Sri Aurobindo, will be the step from man to the Gnostic Superman. We must prepare ourselves for this epochal transformation and the advent of a divine life upon the Earth.

Part and parcel of Sri Aurobindo's evolutionary vision was the absolute necessity of a greater seeing and harmonization that might resolve the confusion and discord of clashing mental ideas that have been imposed upon the world in the service of the collective ego and its insatiable appetites. The catalyst and incentive for this greater seeing, said Sri Aurobindo, would come in the form of a painful evolutionary crisis in which humanity found itself arrested and bewildered and no longer able to find its way. Today we find ourselves in the throes of just such a crisis and imprisoned by conflicting political, religious and economic ideas, in the service of which men are willing to oppress, exploit, and even kill their fellow man. What is abundantly clear is that the multitude of problems we face in the world today cannot be solved by the consciousness that created them. We must seek a higher, more inclusive and enlightened vision if we are to survive these perilous times. But, in order to do this we must first understand the profound limitations of our present Mental species.

The Worlds of Shadows & Light

For thousands of years the poise of human consciousness has been founded on three rather than four pillars of being; the Physical and Emotional in the service of the Mental. The highest, the Spiritual was either lacking or dormant in nearly all of humanity. But for those few in whom the Spiritual was an active principle, the message was always the same: ***'One alone exists, but the sages call it by many names'****(Rig Veda 1-164-46)*. Those lacking that supreme consciousness of Unity were said to be asleep or wandering in the world of illusion. Perhaps the best description of this unfortunate condition, certainly the most iconic, was Plato's 'Allegory of the Cave'. In a dialogue between Socrates and his pupil Glaucon, Plato likens the perception of the ordinary Mental man to a prisoner, chained deep within a cave and facing a wall upon which play the images of light and shadow from a fire burning behind them in the distance. The men and women, chained there since birth know only the shadows and perceive them as the only and absolute reality. According to Plato the shadowy world that the cave dwellers perceive is an illusion of the mind of dualism beyond which lies another, truer Reality. Owing to this, said Plato, all empirical investigation and knowledge amounts in the end only to a great deception grounded in the nature of our Mental faculties and Reason. To perceive the real, the prisoner must throw off his chains and walk out of the cave of the mind into the light of day.

What Plato, Sri Aurobindo and others have affirmed is that there is an epistemological spectrum ranging from the ordinary human consciousness to a supreme enlightenment. The stages of this spectrum proceed from gross to subtle to causal; from the Mental, Higher Mental, Intuitive, to the Overmental and Supramental planes. The lower levels of this spectrum are based upon the reductive intellect and its empirical observations with the higher being gradations of a non-dual consciousness beyond the relativity of the knower and the known. It is a direct experience in which we become one with the truth, one with the object of knowledge. At these higher Supra-mental ranges, one always cognizes essence, truth. There is not even a trace of false knowledge and this is what lies beyond the shadows of Plato's cave.

> *'...Truth cannot be attained by the mind's thought but only by identity and silent vision. Truth lives in the calm wordless Light of the eternal spaces; she does not intervene in the noise and cackle of logical debate.'*[5] **Sri Aurobindo , *Essays Divine and Human*. P. 68**

> *'The nature of Supermind is that all of its knowledge is a knowledge by identity and oneness... The Spirit is one everywhere and always. It knows all things as itself and in itself and therefore knows them intimately, completely, in their reality as well as their appearance, in their truth, their law, the entire spirit and sense and figure of their nature and their working. When it sees anything as an object of knowledge, it yet sees it as itself and in itself, and not as a thing other than or divided from it about which therefore it would at first be ignorant of the nature, constitution and workings and have to learn about them, as the mind is at first ignorant of its object and has to learn about it because the mind is separated from its object and regards and senses and meets it as something other than itself and external to its own being...'*[6] ***Yoga of Self-Perfection by Sri Aurobindo Chapter XIX, The Nature of Supermind, Arya - July 1920***

The Mystery of the Veda

As we look back over our collective history the traces of people with this higher knowledge can be found in all cultures and all times and their teachings form what the writer Aldous Huxley called a kind of 'Perennial Mystical Philosophy'. But it is clear that this advanced realization was confined to a relative few, it was not spread in the whole mass of humanity. Yet there was one ancient civilization whose enigmatic scriptures indicate that they were in full possession of this knowledge and that it formed the basis of their entire cultural experience. While much of the history of the Vedic world remains an inscrutable mystery, their Seers and Rishis preserved the essence of this supreme knowledge and the means of its attainment for future humanity in a compilation of scriptures known as the Veda - Rig, Saman, Yajur and Atharva.

> *'The Vedas, after all, were written by people who remembered a radical experience, which must have taken place on earth at a given moment, as an example of what was to come. (This always happens in the yoga: a first radical experience comes like a herald of the future realization.) So in the terrestrial yoga – in the yoga of the earth, of the planet earth – there was a moment when it came; they who are called the forefathers must have created, through their effort and their yoga, at least an image of the supramental realization. And those who wrote the Vedas, who composed all these hymns, remembered or kept the tradition of that experience. And oh, mon petit, it had the same effect on me as when I read the 'Yoga of Self-Perfection' in The Synthesis of Yoga (Mother catches her breath): <u>there is such a gulf between what we are, what life on earth and human consciousness now are, even among the most enlightened, the most advanced, and THAT!</u>'*[7] **The Mother's Agenda – April 7, 1961**
> **[underlining emphasis mine]**

Although we are constantly being given evidence of the extraordinary advancements of ancient civilizations like the Vedic, Greek, and Egyptian (their spirituality, engineering, architecture, and cosmology), it is almost impossible for someone living in our modern scientific culture with its intellectual sophistication and dazzling technology to imagine that the ancients were, in so many ways, more advanced than we are today. But the fact that we find ourselves in an untenable evolutionary crisis suggests that something in our modern world is dangerously out of balance and in order to avoid a global catastrophe of epic proportions, we must quickly discover why. Our pursuit of these truths requires us to go beyond the glib analyses of the media pundits and into the deep existential issues that color all other aspects of our lives. The first and most important question is: What is the true nature of the Self? The second question, intimately related to the first is: What is the nature of our circumscribing reality and does it have any meaning or purpose? The answer to the first question is a matter for men and women of wisdom, the philosophers, sages and yogis of our time. The answer to the second falls within the purview of the modern physical sciences. Whether they can answer or not is the focus of this study.

One of the most cogent critics of modern science was the late Vaclav Havel, essayist, poet, dissident and former president of the Czech Republic. In a highly acclaimed speech at Independence Hall in Philadelphia, Havel attributed the chaos of the post-modern world to the failure of rationalism and the scientific ethic:

> *'...Today many things indicate that we are going through a transitional period, when it seems that something is on the way out and something else is painfully being born. It is as if something were crumbling, decaying, and exhausting itself, while*

> *something else, still indistinct, were arising from the rubble... The world modern science has fostered and shaped now appears to have exhausted its potential. It is increasingly clear that something is missing. [Science] fails to connect [us] with the most intrinsic nature of reality and with natural human experience...and it is now more of a source of disintegration and doubt than a source of integration and meaning... Today, for instance, we know immeasurably more about the universe than our ancestors did, and yet, it increasingly seems they knew something more essential about it than we do, something that escapes us.* '[8] Vaclav Havel, **Independence Hall Speech, Philadelphia, July 4, 1994.** [underlining emphasis mine]

According to Sri Aurobindo and the members of his Line, '...that something essential that escapes us' is the secret concealed in the Veda. When it is entirely discovered, said Sri Aurobindo, '*...it will be found to formulate perfectly that knowledge and practice of divine life to which the march of humanity, after long wanderings in the satisfaction of the intellect and senses, must inevitably return*'. But the realization of these eternal truths requires more than just a scholarly, intellectual knowledge of the Veda. To the uninitiated, these ancient hymns appear to be the remnants of a naturalistic religion that celebrated the materialistic desires of a barbaric mind and life. They remain unintelligible to scholars and pundits who have not made that inner plunge by which they might recognize the symbols of the Rishis as revelations of the supraphysical action, forces and processes which created the physical world. Since it is by those same forces and processes that our inner life is governed, knowledge of the Veda can only be recovered by the same means it was originally possessed - through the lived yogic experience. Because of this limitation, wrote Sri Aurobindo, *'...for some two thousand years at least, no Indian has really understood the Vedas'*[9].

Since one of the greatest mysteries of the Veda is Time, it should come as no surprise that the supreme knowledge contained in its hymns has been preserved through the ages by Time itself, as if in a Seed waiting for the precise moment of its unfoldment. It is equally no surprise that there would be attending its timely unveiling a group of individuals with a unique alignment of consciousness who could recognize and reveal its precious contents. There have been three such individuals whose writings on the Veda stand unequalled by any other known thinkers: Sri Aurobindo, his collaborator in the Supramental Yoga called the Mother, and Patrizia Norelli-Bachelet, known to her students throughout the world as Thea. Sri Aurobindo and the Mother each with their own unique gifts have come and gone. Thea remains today continuing their epochal work and revealing ever deeper mysteries of Vedic Time and Cosmology. It is at Thea's level of the work that we discover a new knowledge beyond both science and spirituality, one which promises to reconcile them both into a fully integrated synthesis which will allow for a collective experience different from anything the world has ever known. Since the proof is in the pudding so to speak, we must examine Thea's discoveries in the light of modern scientific theory, a comparison which will reveal a depth of knowledge that our most celebrated physicists have been unable to achieve.

A Brief Overview of Science

Any review of the history of modern science must naturally begin with the 16th century astronomer, Nicolaus Copernicus. In 1543 Copernicus published his *De Revolutionibus Orbium*

Coelestium, a book which proposed a radical new heliocentric model of the universe placing the Sun, not the Earth, as the center of the Solar System. While his cosmological theories were considered revolutionary at the time, details of the heliocentric universe had been discovered in India some 5,000 years earlier and were included in the sublime hymns of the Rig Veda. A number of these ancient hymns explain both the physical and metaphysical characteristics of the Earth and Sun and have been interpreted as stating that the Sun is at the center of the solar system and that the Earth circled the Sun. This heliocentric knowledge is perfected by the Vedic sage Yajnavalkya (9th–8th century BCE) in his astronomical text, the *Shatapatha Brahmana*. Due to the absence of Vedic knowledge in the ancient and medieval world, Copernicus' discoveries were accepted as unique and original. They challenged the cosmological authority of Ptolemy and Aristotle and caused a complete scientific and philosophical upheaval that ended an old era of understanding and began a new scientific revolution. Johannes Kepler was born some 98 years after Copernicus and was able to build on his revolutionary discoveries by proving that the planets moved in an elliptical orbit around the Sun. Sir Isaac Newton arrived 74 years later and used Kepler's laws as a basis of discovering how the laws of gravity caused the Earth and other planets to follow that elliptical path. The work of these extraordinary individuals ignited a renaissance in physics that became the basis of a new science which lasted for nearly 200 years.

Einstein & the Cosmological Constant

1905 marked the beginning of the end of the Newtonian era when Albert Einstein appeared on the scene. He was a deeply spiritual man driven by an insatiable curiosity and a desire to formulate, as he put it, one simplified and lucid image of the world. Unlike many of his fellow scientists, Einstein never accepted that the universe was simply a random creation. This belief was expressed in his comment that *'God does not play dice'*. His confidence in the eventual discovery of a supreme unifying order led him to say that the goal of science is nothing else but to discover the mind of God. In 1905, Einstein rocked the world of physics with the publication of his Special Theory of Relativity which proposed that energy and matter are linked together by the now famous equation: $E=MC^2$. In this equation, E represents energy, M represents mass and C^2 is the square of the speed of light. Einstein's theories claim that the speed of light in empty space is a natural constant the value of which never changes, and that nothing can exceed the speed of light. His new discoveries revealed that Newton's Laws of Motion, which had been considered sacrosanct for over a hundred years, were only approximations that broke down when velocities began to approach the speed of light. In 1915 Einstein published the General Theory of Relativity which showed that Newton's Law of Gravitation was also only approximately correct. He agreed that Newton's laws were still valid for smaller masses but did not work for extremely large bodies. According to Einstein's General theory of Relativity, massive objects curve spacetime; the larger the mass, the more spacetime is curved. His General Theory of Relativity completely changed the way that astronomers look at the universe to include the influence of black holes and how light from distant stars is distorted by gravity. It provided equations that could explain everything from the motion of the planets to the trajectory of light with incredible accuracy.

But as extraordinary as his theories were, Einstein discovered that his equations didn't quite work for a static universe. The inexorable force of Gravity would cause a universe which was initially at rest to collapse under its own weight, so he calculated a hypothetical repulsive force, a principle of anti-gravity that would fix the problem and balance things out at every point in space. He called this unknown force the 'Cosmological Constant' saying that it would stabilize the universe against the attractive effect of gravity. In a 1917 paper entitled *Cosmological Considerations in the General Theory of Relativity*[10] Einstein added an extra term to the equations of general relativity. The Greek capital letter lambda 'Λ' was proposed by Einstein as a modification of his original theory of general relativity to achieve a stationary universe and to preserve the symmetry and elegance of his mathematical formulas. It now read: E = MC^2 .The new formula reflected the addition of a force that could counterbalance gravity, maintain an equilibrium of forces and prop up the universe. Einstein's constant existed everywhere throughout the vacuum of space as a kind of latent energy. It was described as a 'Nothing that is Something'. And while he later had to disown it due to the groundbreaking discoveries of the astronomer Edwin Hubble, Einstein had unknowingly stumbled upon one of the greatest mysteries of creation.

While working at the Carnegie Observatories in 1929, Edwin Hubble began to measure the red shifts of a number of distant galaxies. 'Red shift' is a key concept for astronomers which measures light moving toward the red part of the spectrum. It is very much like the 'Doppler effect' which measures the changing pitch of sound waves as they approach and pass an observer. Just as there is an audible change in pitch as the sound source approaches and passes a fixed point, there is a similar change in the wavelength of light indicating whether its source is moving toward or away from us. Light coming from distant objects that shifts toward the red end of the spectrum is an indication of an increasing distance to the object due to the expansion of space itself. When Hubble measured the relative distance of these galaxies and calculated the redshift of certain stars called 'Cephids' in each galaxy, he concluded that the only possible explanation for the data he was receiving was that the universe was rapidly expanding and moreover that it must have been expanding since the very moment of its creation. Once Hubble had discovered that the universe was expanding and apparently not in a 'steady-state' of equilibrium, Einstein immediately discarded the cosmological constant, declaring it to be the biggest blunder of his life. Little did he know that his constant would be revived decades later in an attempt by physicists to explain new theoretical concepts known as the Quantum Vacuum and Dark Energy.

David Bohm & The Nasadiya Sukta

Following Hubble's astronomical discoveries of an expanding universe, scientists did the math and realized that if the universe were currently expanding, it must have been much smaller in the distant past. Their extrapolations indicated that at some earlier moment the entire universe must have been compressed into a single point or 'singularity' no larger than a sugar cube. The implications of this deduction were extraordinary because they posited a hypothetical starting point for the physical creation itself as well as the need to understand that indescribable 'something' from which the universe sprang into being. The notion of a cosmic Singularity led to the development of an entirely new scientific paradigm based on the belief that the molecular and atomic properties of matter in a more dense quantum era might provide important clues for deriving its macrocosmic properties. But unbeknownst to Quantum Physics, as this new discipline was to become known, the Vedic seers had made this discovery many thousands of years before. According to the Rig Veda X. 149 '*...a universe comes to birth from its center; it spreads out from a central point that is as it were, its navel. It is in this way that the universe was born and developed from a core, a central point'*. Moreover there is a very specific point in the spatial plane where this break occurs. In the ancient Vedic culture this singularity or point inspired the symbolism of a Cosmic Navel or Axis Mundi which was situated at the center of the world and constituted an inviolable link between heaven and earth. Their holy sites and cities were built on the basis of this realization as well as the architectural geometry of their most sacred temples.

With Einstein's and Hubble's discoveries, the stage had been set for new explorations into the nature of matter and the dynamics between the subtle and physical universes. For it was now becoming accepted doctrine that the origins of the material universe lay either at the threshold of the Quantum Singularity where everything was undifferentiated and squeezed into a tiny point or utterly beyond in a realm which they could neither observe nor measure. Scientists were slowly beginning to understand what the Vedic Rishis had known for millennia.

> *'The material universe is only the facade of an immense building which has other structures behind it, and it is only if one knows the whole that one can have some knowledge of the truth of the material universe. There are vital, mental and spiritual ranges behind which give the material its significance...I myself see certain fundamental truths underlying all the domains and the one Reality everywhere. But there is also a great difference in the instruments used and the ways of research followed by the seekers in these different ways (the physical, the occult and the spiritual) and for the intellect at least the bridge between them has still to be built. One can point out analogies, but be maintained very*

well that Science cannot be used for yielding or buttressing results of spiritual knowledge.' [11] ***Sri Aurobindo, Letters on Yoga,*** **[underlining emphasis mine]**

Of all the legendary figures associated with the early days of Quantum Physics there was one man who exhibited a unique grasp of the true nature of Reality and the multiple subtle dimensions that lay behind the physical. David Bohm was an American physicist and author of one of the most impressive theories emerging out of scientific cosmology respecting the ancient Vedic truths. In his book, *Wholeness and the Implicate Order* Bohm takes us behind the facade of the material world and attempts to describe the transcendent reality as a graded energetic hierarchy representing energy in successive states of manifestation from infinitely subtle to the gross physical reality.

> *'...The Explicate Order, weakest of all energy systems, resonates out of and is an expression of an infinitely more powerful order of energy called the Implicate order. It is the precursor of the Explicate, the dreamlike vision or the ideal presentation of that which is to become manifest as a physical object. The Implicate order implies within it all physical universes. However, it resonates from an energy field which is yet greater, the realm of pure potential. It is pure potential because nothing is implied within it; implications form in the implicate order and then express themselves in the explicate order. Bohm goes on to postulate a final state of infinite [zero point] energy which he calls the realm of insight intelligence. The creative process springs from this realm. Energy is generated there, gathers its pure potential, and implies within its eventual expression as the explicate order.'* [12] **Will Keepin, *David Bohm*, Noetic Science Journal**

While Bohm posited the Implicate order as the primary dimension of energy he was unable to provide direct knowledge of its existence. It could only be intuited through observations and experiments on the secondary dimension, the Explicate order. This earned him some sharp criticism from his fellow physicists who thought that empirical observation of the known was the only sure course to discover how the universe works. But with recent discoveries of so-called 'dark matter' and 'dark energy' it appears that no less that 96% of the universe is completely unknown to us and more importantly, unmeasurable. A reductive empirical observation of the known is simply no longer an option. If scientists can only guess what is going on based on the gravitational effects on the 4% of the universe that they can measure, this is clearly insufficient to provide any credible picture of the universe, let alone a grand unified field theory.

In spite of the controversy, Bohm's theories are sure to be validated in many particulars because they follow a solid metaphysical premise. In many ways they sound like they have been lifted right out of the Rig Veda which describes a similar arrangement of consciousness. His realm of pure potential bears an amazing resemblance to an original matrix of consciousness known in Sanskrit as *'mahat brahman'*. The mahat is on the one hand the un-manifested form of matter and on the other hand pure consciousness. It can be properly thought of as the matter-consciousness-field. To appreciate the nature of this primal mystery we must turn to the 129th hymn of the 10th Mandala of the Rig Veda where we will find the 'Nasadiya Sukta' also known as the 'Hymn of Creation'.

In this ancient hymn the Rishi sings of a reality whose Oneness is so profound that every single relative distinction is overcome producing an unutterable and unthinkable awareness that can only be described as 'not this, not that'. 'Neti-neti' is not a negation or denial, but an expression of an experience of the

Absolute which the human mind can only register as inconceivable. All attempts to describe it end up being a fragmented corruption of the whole. Because of this irreducible wholeness, the Rishi cannot tell us what it is, so he introduces us to the great mystery through a series of questions.

Nasadiya Sukta:

'1. At first was neither Being nor Nonbeing.
There was not air nor yet sky beyond.
What was its wrapping? Where? In whose protection?
Was Water there, unfathomable and deep?
2. There was no death then, nor yet deathlessness;
of night or day there was not any sign.
The One breathed without breath, by its own impulse.
Other than that was nothing else at all.
3. Darkness was there, all wrapped around by darkness,
and all was Water indiscriminate. Then
that which was hidden by the Void, that One, emerging,
stirring, through power of Ardor, came to be.
4. In the beginning Love arose,
which was the primal germ cell of the mind.
The Seers, searching in their hearts with wisdom,
discovered the connection of Being in Nonbeing.
5. A crosswise line cut Being from Nonbeing.
What was described above it, what below?
Bearers of seed there were and mighty forces,
thrust from below and forward move above.
6. Who really knows? Who can presume to tell it?
Whence was it born? Whence issued this creation?
Even the Gods came after its emergence.
Then who can tell from whence it came to be?
7. That out of which creation has arisen,
whether it held it firm or it did not,
He who surveys it in the highest heaven,
He surely knows or maybe He does not! '

[13] **Rig Veda X, 129**

The Hymn of Creation is an incomparable statement of what lies beyond the Quantum Singularity. It clearly describes a boundary, a *'crosswise line'* separating Being from Non-Being. *'...Bearers of Seed there were and mighty forces...'* sings the Rishi, expounding symbols of what we may presume gave birth to the universe. To do full justice to this hymn would require a lifetime of study and volumes of commentary but for the purposes of this study we need only consider one thing, the hymn succeeds in describing the unutterable Fullness of the Transcendent. If there is one mantra repeated throughout all of India's ancient scripture it is, *'From fullness is born fullness. When fullness is taken away from fullness, Fullness still remains'*(Brihadaranyaka Upanishad, 1.10).This is the greatest mystery and the reason why the Vedic and no other religion or philosophical system on Earth can rise to the task of producing a fully integrated synthesis of knowledge.

For millennia spiritual seekers of all persuasions have scaled the heights of consciousness to unmask the mysteries contained in this primordial hymn. Many have succeeded in dissolving their small 'self' in that larger Self whose essence is Brahman, the non-dual absolute. Most of these, said Sri Aurobindo, *'...arrive at their object, but only to fall asleep in the infinite'*[14]. Of those who have succeeded none, until now, have been able to build a bridge linking the Transcendent and the luminous planes of consciousness that exist between THAT and the mental human being. But it is these intervening planes that contain the alchemical mystery of how Consciousness becomes Matter; how That becomes This. To date, only Sri Aurobindo and his line have been able to fully articulate the spectrum between the Absolute and the Material world. And while Bohm's theories of the Implicate and Explicate orders are extremely helpful in understanding the holism of the universe, he was never able to appreciate the role of Time as the integrating principle that joins these two orders of existence. None but Thea (Patrizia Norelli-Bachelet) has been able to give out the complete details of a Supramental Cosmology that describes the occult processes that occur at the nexus of the Matter-Consciousness fields.

Time: The Supreme Mystery

In the Nasadiya Sukta the Rishi introduces us to the Transcendent, an unthinkable concept that may only be known through the direct lived experience of the extra-cosmic Absolute. It is an awareness of consciousness extending to the ultimate reaches of itself where all relatives in the universe dissolve into a great and immense calm. In this sublime condition, there is no movement because there is no Time. It has often been described by the mystics as an 'eternal now of Being'. But slowly That, emerging out of its immobility, stirring through its own power crosses an invisible boundary where it links up with Time and Space and begins to unfold linearly as the Becoming. All that had existed in the womb of Being, though unmanifest, could now unfold in Time and extend itself in Space. In her book, *'Questions and Answers'*, the Mother, one of the greatest spiritual realizers of our age, gives us a first-hand account of that experience.

> *'...One movement is within, containing everything in itself, without any expression of what is there; and the other movement is just this [the Mother makes a gesture of projection], and all that is within oneself comes out.*
>
> *And then, for this to become perceptible, it must be continuous. When it is within, it can be simultaneous, for it is un-manifest, so all is in an eternity outside Time and Space – immobile, inexistent. In the opposite direction, everything*

becomes and so there is a continuity of perceptions which follow one after another and spread out in Space and Time.

And it is the same thing.

It is exactly as if you are like this [the Mother makes a gesture of being doubled up], and then you do this [open up] – and so what was there comes out. So these two movements are literally opposite, but it is the same thing in two opposite attitudes which are simultaneous: it remains like this [inward gesture], and at the same time it is like this [outward gesture]; the one does not cancel out the other and they exist simultaneously. But in one direction it is imperceptible because it is contained within itself, in the other movement it is thrown outside, and so it can be seen. And when it is self-contained, it is co-existent in a perfect simultaneity; and in the other movement it unfolds itself in a constant becoming. And when it unfolds itself, it necessarily creates Time and Space, while there it is outside Space, outside Time and beyond all possible perception. But it is the same thing in two opposite movements.

And that is what truly is.

It is like that.... One single thing in two opposite aspects. '[15]

The Mother - Questions and Answers, 11 April 1956, (p. 108)

'*...And when it unfolds itself, it necessarily creates Time and Space*' says the Mother in such a matter of fact way that you could almost miss the fact that she has just described one of the most profound mysteries of human existence and the Holy Grail of Quantum Physics. In this brief statement the Mother has given us a glimpse into the nature of the relations between the Spiritual and Physical worlds and the 'mechanism' by which one becomes the other. What we have not understood, until now, are the dynamics attendant to the creation of Time and

Space and the nature of the interaction between these two fundamental principles that underlie all aspects of the physical universe. In order to pierce this veil we must return to the font of Indian wisdom. In his commentaries on the *Brihadaranyaka Upanishad,* one of India's oldest and most profound metaphysical texts, Sri Aurobindo discusses the occult imagery associated with these fundamental principles and the nature of their powers:

> *'... The [image of the] horse is a physical figure representing like an algebraical symbol, an unknown quantity of force and speed. (Time is its breath, the year is its body, the seasons its limbs, the months and the fortnights its joints, the days and nights its feet.) From the imagery it is evident that this force, this speed, is something universal. Time in its period is the Self of the Horse Sacrificial, so not Matter but Time, is the body of this force of the material universe... Space then, is the flesh constituting materially this body of Time which the sage attributes to his Horse of the worlds, by movement in Space its periods are shaped and determined. Hence the real power, the fundamental greatness of the Horse is not the material world, not the magnitudes of Space, but the magnitudes of Time... for Time is that mysterious condition of universal mind which alone makes the ordering of the universe in Space possible.* '[16] **SriAurobindo, *The Upanishads; 'The Great Aranyaka'*, Sri Aurobindo Ashram Press, 1971 [underlining emphasis mine]**

While it may seem counter intuitive to the non-initiate, what Sri Aurobindo tells us in this extraordinary passage is that it is Time that creates Space, not the other way around. The preeminence of Time has been acknowledged in Indian iconography for millennia through the symbol of Shakti, the divine cosmic force, dancing upon the inert body of Shiva, with the serpent of eternal time wrapped around her body. But to properly understand Time's ordering function we must go beyond the notion that Spirit and Matter are two separate things. They are in reality a continuum of consciousness in different states of vibration. The Spiritual plenum is characterized by a complete fusion of Time and Consciousness which is experienced as Timelessness, or undifferentiated Time. In this state Time-energy vibrates at such an intense rate that the laws of physics break down and scientists are unable to measure its infinite movement. The closest they have been able to come to that Zero threshold is the first millionths of a second after the so-called Big Bang, a measurement known as 'Planck time', named after the physicist Max Planck. Their inability to go behind the veil and reproduce measurable conditions at this infinite threshold has led many physicists like Stephen Hawking to a deeply erroneous conclusion.

> *'...If there were events before the 'Big Bang', one could not use them to determine what would happen afterward because predictability would break down at the 'Big Bang'. As far as we are concerned, events before the 'Big Bang' can have no consequences and should not form part of a scientific model of the universe.'* [17]***Stephen Hawking – A Brief History of Time.***
> **[underlining emphasis mine]**

Did Hawking actually say that? What right has he to conclude that processes which lie beyond science's capacity to measure are inconsequential? This is a perfect example of the hubris of modern science and the limitations of the empirical method. And it has led to a fundamental error that calls into question the entire premise of theoretical physics**: that there are three dimensions of Space and only one dimension of Time**. The entire problem of modern science is that no matter how close they seem to come to the truth, they are hopelessly blocked by the form of their inquiry. One of their own, the Nobel laureate Werner Heisenburg, explains: '...*What we are observing is not nature itself, but nature exposed to our method of questioning*.'[18] And this has led scientists like Hawking to the unfortunate error of mistaking the content of their thought for an accurate description of the world. As Patrizia Norelli-Bachelet so wisely concludes:

> '...*The difficulty with scientific observation is that each time the point is reached when the truth could be known, it becomes necessary for the scientist to enter another dimension for knowledge,—the realm of consciousness. But as its existence is denied, the truth recedes ever and always, just as the core of the universe recedes from science's probing vision and the heart of Matter is never really known.*'
> [19] **Patrizia Norelli-Bachelet, *The New Way, (pp. 260-261)***

What is demonstrably lacking in the scientific quest is a knowledge obtained through a direct experience of the fundaments of reality and this must be done like the Rishis did it, like Sri Aurobindo, the Mother and Patrizia Norelli-Bachelet have done, not in the content of their thought, but through a process of identity in consciousness that gives rise to a science of the whole of reality and a non-speculative description of its order.

Vedic Cosmology

The only place in the world today where one can explore the philosophy of the Vedic seers and experience the living principles behind their teachings is the Aeon Centre of Cosmology in south India. Aeon Centre is a microcosmic laboratory where its director Patrizia Norelli-Bachelet observes and records for publication the direct collaboration of Material

Nature acting to unveil a new principle of order in the world. In addition to applying the principles of this New Cosmology to the pressing questions of modern science, Ms. Norelli-Bachelet is able to demonstrate these irreducible principles in a variety of other fields including yoga, mythology, history, sacred architecture, geopolitics and music. But one of the most interesting fields of her Applied Cosmology, one that respects the Upanishadic principles of Force, Speed and Time that Sri Aurobindo wrote about, is her long time passion for breeding and racing thoroughbred horses. In fact, it is the horse that has provided some of the most profound insights into new applications for her work. As we go deeper into the formulas of this New Cosmology it is important to understand that they are not the same stuff as the arcane mathematics scrawled across the chalky blackboards of Ivy League universities. The formulas of the Supermind are living, breathing realities that can be observed by anyone in the ordinary passage of a human life. As the physicist-chemist Dr. Peter Plichta confirms:

> *'...God arranged the world simply. He certainly did not intend that comprehension of His world would be restricted to elite groups in universities.'* [20] ***God's Secret Formula - Deciphering the Riddle of the Universe and the Prime Number Code***

The simplicity of Ms. Norelli-Bachelet's Cosmology lies in its ability to describe the enfolding and unfolding of the universe through the mechanics of an ordinary seed; the very same seed the Rishi sings about in the Creation Hymn. It follows the same organic process which allows a tiny acorn to become the mighty Oak. The Oak is already there enfolded in the acorn, there in Being, but it is the power of compressed Time that is the driver of its Becoming. There is a wonderful description of this seeding process in the Hiraṇyagarbha Sūkta of the Rig Veda[21] as well as the Matsya Purāṇa. The seer begins by saying: '*...there was darkness everywhere. Everything was in a state of sleep. There was nothing, either moving or static. Then the 'self-manifested' or 'that which is created by its own accord' arose. It created the primordial waters first and established the seed of creation into it. The seed turned into a golden womb, Hiraṇyagarbha, which is the source of the entire creation.*' [22]

According to Ms. Norelli-Bachelet, the occult dynamics of this creation process can be reduced to two primary forces or what she calls the two 'cosmic directions': Contraction and Expansion. Contraction is the property or 'direction' of Time and Expansion is the 'direction' of Space. Contraction is on 'the other side' of the singularity and it is constant. It is in fact the same cosmological constant that Einstein disowned as the 'biggest blunder of (his) life'. It is not an 'energy' as Einstein originally believed but rather an 'all pervading cosmic direction' which, by virtue of its contracting action, counterbalances the expansion that Hubble noted on this side. Unbeknownst to contemporary physicists, this is the origin of Gravity. But to understand the nature of the gravitational force one must know Time, and that knowledge lies on 'the other side' of the big bang hypothesis. Of course Stephen Hawking has authoritatively told us that whatever happened before the big bang '*...can have no consequences and should not form part of a scientific model of the universe.*'

The picture that emerges from Ms. Norelli-Bachelet's cosmology is of a creation that proceeds out of an incalculable Infinite, the nature of which is energy-consciousness in a completely undifferentiated state. This is the vast Transcendent which, in the act of manifestation and by its own accord, is reduced to a seed of itself, a Zero Womb. The compressed energy within this seed becomes ordered as a triadic force known as the three qualitative modes of nature or 'Gunas' that are intertwined in all cosmic existence. The 'Gunas' are an ordering principle which all things must obey from literally the first moment of creation. In Vedic metaphysics the gunas are called *Rajas, Sattwa* and *Tamas*, energy flows which carry the functions of Creation, Preservation and Dissolution.

The elegant simplicity that exists in that Zero Womb, says Ms. Norelli-Bachelet, is reflected in whole Numbers because numbers alone can explain that reality.

> *'...There Number IS, whole and complete in itself. Physics, math as we know them cannot explain – much less prove – the elegant simplicity of that Unity. Whole numbers, as number-powers, alone 'explain' reality...The Zero/Contraction is what holds the universe together. It is the binding force – the eternal all-pervading Centre. Without that there is no universe at all. And those number-powers encapsulate the Transcendent, the Cosmic, the Individual Soul. These are the triadic principles of all creation, of all created things. These principles through the 0/1 are at the root of our universal manifestation – and they are the components of the individual Soul, our vahana in this material creation of 9.'*[23] ***PatriziaNorelli-Bachelet, 'The Miracle of Oneness', 2012, p.1***

The primacy of Numbers is not a new concept. Pythagoras and Plato had long proclaimed that numbers contained the truth of all things, that indeed, numbers rule the universe. We can see this primordial truth expressed in the three gunas which possess a numerical equivalence to the numbers 9, 6, and 3. When this triadic seed of time, born out of contraction, crosses the quantum threshold the direction shifts to expansion, giving rise to the Point of Space. In the Vedic myths, this is known as the birth of the One, the Immanent, 'upholder' of the worlds and 'support' of the universe. The One emerges across the threshold of Time as its fourth dimension. Thus the formula of Ms. Norelli-Bachelet's Supramental Cosmology is **9-6-3-0/1**, a simple number sequence that describes the process by which solidified energy, form and matter, emerge from that zero womb to begin its orderly evolution in Space. **It is a formula born of three dimensions of Time and One of Space which is a complete reversal of the present understanding held by modern physics.** This is precisely why science cannot discover the underlying order and organization in the Cosmos, it hinges on a complete misunderstanding of Time. Ms. Norelli-Bachelet takes us deeper into the workings of this divine alchemy and its ability to order the universe.

> *'Once the Boundless is closed within boundaries, the energies thus enclosed experience a process of ordering. The most important feature of this process of ordering is the emergence of a 'center', out of which arises an 'axis'. And, when a certain threshold is reached, it is this axial alignment which can hold or gather systems to itself, be this a Cosmic, Universal, or Solar system or a nation, race, or individual.'* [24] **Patrizia Norelli-Bachelet. *'Supermind and the Language of Gnostic Symbols', The Vishaal Newsletter, Vol. 9/3, August 1994,***

> *'Science must then ask what the nature of an axis is and what are its origins. We are of course aware that macrocosmic bodies do indeed possess axes and that they 'churn away' around their fixed poles. But how did this come into being in the first place? We seem to take for granted that axes always existed or that in some way, unknown to us, they arose simultaneously with the first crystallised substance. This hypothesis may not be sufficient to understand the nature of our physical dimension because as the ancient Vedic Seer*

> *realised in the dawn of history, the discovery of the origin of an axis is essential both for the proper understanding of the cosmic process and evolution, as well as the evolution of smaller bodies and finally the human species. Given this central importance, the axis is found in all cultural expressions which can trace their roots to that original Vedic* act of seeing.'
> [25] **Patrizia Norelli-Bachelet.** ***Culture and Cosmos, The Vishaal Newsletter, Vol 8 /2, pp. 20-21 June 1993,***

In these revelatory passages Ms. Norelli-Bachelet gives us a glimpse into the core of creation. When the energies of the Zero womb undergo a process of compaction to a point, an 'Axis' arises which can '*hold or gather systems to itself*' to form an ordered cosmos. This principle of axial alignment is self-evident in a wide range of ordered correspondences from Galactic, Solar and Planetary systems, descending down to the individual. The importance of this series of equivalencies is that the self same structure and order of the Macrocosm may be discovered at many intervening levels before finding a highly specific individuation in the Microcosm. But it is only at the individual level where one may actually experience these eternal patterns and cycles through the realization of a common center. Since the same Zero Womb, which gave birth to the universe is also the core of the human being, the unveiling of this Soul/center/axis in the individual gives rise to a simultaneous realization of Truth-Consciousness or *Gnosis*.

Indeed, the Cosmic manifestation, from its highest plane down to physical nature, is structured in such a way as to reflect a single interrelated whole expressing one being, one life, and one consciousness evolving toward a divine apotheosis. It is patterned in a way so that the same movement, order and goal may be seen in the life of the individual as well as in the entire body of the manifest universe. The realization of this multidimensional whole and its sacred and eternal order is the supreme goal of the Vedas because it endows one with the keys for the organization of all human knowledge. Sri Aurobindo's Supramental Yoga is based upon this realization and promises the seeker a Truth far beyond the knowledge of Science and its reductive Mental paradigm.

The Emperors Have No Clothes

In a statement reminiscent of the images in Plato's Allegory, the American poet and philosopher Ezra Pound wrote, *'Our science is from the watching of shadows.'*[26]But was it always this way? Perhaps some of you share my memories of high school science class where we learned about Galileo, Louis Pasteur, Madame Curie, and Jonas Salk, iconic scientists peering through their telescopes or microscopes, huddled over their beakers and Bunsen burners in the act of analyzing and measuring the important properties of tangible, physical matter. This is a very different image from the captains of modern science who need little more than a laptop computer to ply their trade. While they claim to be scientists in the classical sense of the word, the truth is that many of them have no right to the title. With the advent of Quantum Physics many scientists are unable to test their exotic theories because the immense densities, pressures and temperatures which lie at the quantum singularity cannot be reproduced under laboratory conditions. Since they cannot actually 'test' and prove their theories, many of these physicists have been reduced to the status of philosophers. What used to pass for reliable knowledge is now no more than an educated guess, and even that is subject to question given the limitations of the mental consciousness and the scope of its inquiry. Albert Einstein's theories of Special and General Relativity suffer this same limitation because they attempt to describe a reality beyond the limits of the domain which physics is competent to measure; a reality in which the Transcendent can SEE a truth, knowable to It alone.

> '...*Einstein's law is a scientific generalisation based upon certain relations proper to the domain of physics and, if valid, valid there in the limits of that domain, or, if you like, in the general domain of scientific observation and measurement of physical processes and motions, but how can you transform that at once into a metaphysical generalisation? It is a jump over a considerable gulf - or a forceful transformation of one thing into another, of a limited physical result into an unlimited all-embracing formula. I don't quite know what Einstein's law really amounts to - but does it amount to more than this that our scientific measurements of time and other things are, in the conditions under which they have to be made, relative because subject to the unavoidable drawback of these conditions? What metaphysically follows from that - if anything at all does follow - it is for the metaphysicians, not the scientists to determine...<u>Einstein's views outside his domain are crude and childish, a sort of unsubstantial commonplace idealism without grasp on realities</u>...*'[27] ***Sri Aurobindo, Letters on Yoga,*****[underlining emphasis mine]**

> '...*In the West the higher minds are not turned towards spiritual truth but towards material science. The scope of science is very narrow: it touches only the most exterior part of the physical plane. And even there, what does science really know? It studies the functioning of the laws, edificates theories ever renewed and each time held up as the last word of truth!*'[28] **Sri Aurobindo, In an interview on 8 May 1926, that first appeared in the Sri Aurobindo Circle Eighth Number, 1952, pp. 99-102.**

The result of this persistent overreaching has been the creation of a body of scientific dogma that has taken on the character of a fundamentalist religion where many of its high priests go far beyond their competency to make authoritative statements on matters that they know absolutely nothing about. And the reason that no one challenges their specious conclusions is that almost no one outside their own elite academic circles can understand their complex mathematical formulas. But there was one man who was able to pierce the veil of pretentious calculations and get to the truth. Nikola Tesla was a scientific genius and a profoundly realized man with a deep knowledge of the inner dimensions of reality clearly evident in his statement, '*If you only knew the magnificence of the 3, 6 and 9, then you would have a key to the universe.*' Blessed with this unique perspective Tesla penned a withering critique of Einstein's General Theory of Relativity:

> '*...It is a magnificent mathematical garb which fascinates, dazzles and makes people blind to the underlying errors. The theory is like a beggar clothed in purple whom ignorant people take for a king* .'[29] **Nikola Tesla in an interview with the New York Times in 1935**

Tesla's observations are perhaps truer now than they were in his generation. And while the public remains dazzled by the icons of modern science and accords many of them rock star status, we need only look at their accomplishments, or profound lack thereof, to get a picture of science's true condition. With all the billions spent in scientific research modern physics has a dismal record of being able to come up with answers to some of the most basic questions about our universe. One of the most troubling of these questions deals with the unresolved conflict between the two reigning theories that attempt to explain our reality. Today the universe is described by two partial theories: Einstein's General Theory of Relativity, which details the forces of gravity and large scale structures, and Quantum Mechanics which deals with physical phenomena on extremely small scales. These two theories are known to be inconsistent with each other and have been for over 80 years. With the billions poured into scientific research, what can possibly be keeping Science from arriving at a grand unified theory of Reality other than a fatally flawed approach to the problem? And this is by no means the only example of their failures. I have included a list of important questions that remain unanswered today:

1. What existed before the Big Bang?

2. How Did the Universe come into being?

3. Is there a scientific Explanation for the Universe?

4. Does the Universe Have a Goal or Purpose?

5. Is Evolution Compatible with Creation?

6. Are the Laws of Nature Absolute?

7. What is the Nature of Space and Time?

8. What is Dark energy?

9. What is Dark matter?

10. What is Gravity?

11. How many Dimensions are there?

12. How did Life emerge?

13. What is particle charge and Spin?

14. What is the nature of an Axis and what are its origins?

15. How are certain quantum particles able to transfer information faster than the speed of light?

16. Why is the Anthropic Principle, a theory which holds that the universe was created with precisely the right

properties to support intelligent human life, rejected by Science?

17. Where do the numerous constants of nature come from and how did they emerge?

18. Is there any correspondence between the laws of physics and the laws of consciousness?

19. With all of the evidence pointing to an impeccably purposeful creation, why does Science continue to reject the possible existence of a Transcendent Creator and Sustainer of the Universe?

20. How does science explain the role of time in Jung's theories of synchronicity and Sheldrake's hypothesis of formative causation?

The list goes on and on and why? Instead of concentrating on fundamental questions, scientists prefer to let their theoretical speculations run wild in the pursuit of multiple universes and exotic theories requiring numerous additional dimensions of reality, anything that can produce a 'forced fit' of relativity and quantum theory. The cosmologist Lee Smolin comments on one of the most popular of these hypothetical theories:

> '*…The paradoxical situation of string theory—so much promise, so little fulfillment—is exactly what you get when a lot of highly trained master craftspeople try to do the work of seers.*'[30] **Lee Smolin, *The Life of the Cosmos*, Oxford University Press, 1997**

Since they are unable to pierce into the domain of consciousness for real answers, physicists simply push back the horizon of matter with discoveries of more and more subtle and exotic particles which they parade before the public as evidence that they have almost captured the prize… the elusive Higgs Boson better known as the 'God Particle'.[A] This charade puts one in mind of that wonderful Danish fairy tale written by Hans Christian Anderson called 'The Emperor's New Clothes'.

> *"Once upon a time there was a vain Emperor who was so fond of new clothes* **(read theories)** *that he changed outfits almost every hour and loved to show them off to the public. He spent all his money on appearances. One day two crooked weavers came along and promised the Emperor the newest, finest, best suit of clothes from a fabric invisible to anyone who was too stupid and incompetent to appreciate its quality. 'Besides being invisible, your Highness, this cloth will be woven in colors and patterns created especially for you'. Of course the Emperor cannot see the cloth himself, but pretends that he can for fear of appearing unfit for his position; his ministers do the same. When the weavers report that the suit is finished, they mime dressing him and the Emperor then marches in procession before his subjects, who play along with the pretense. Suddenly, a child in the crowd, too young to understand the desirability of keeping up the pretense, and too honest to ignore the obvious blurts out that the vainglorious Emperor is wearing nothing at all."*[31] **Hans Christian Anderson, 'The Emperor's New Clothes' first published in 1837.**

You could not find a better description for the vain pretense of modern theoretical physics. The Emperors of Science have no clothes, they have no answers and perhaps worse, they have no prospects. What must be clearly understood is that because of the frame of its inquiry and its exclusive dependency upon the empirical method, science has limited itself to only a partial truth, a single dimension, 4% of the total reality that a cosmos contains. Within that domain it is limited to the surface of things, the 'crust' as it were, proffering material explanations for all outcomes. Because of this physicists find themselves obliged to define our universe as an ever increasing abstraction; matter in relation only to itself, without need of anything outside itself to give it law, meaning or purpose, or to serve as its creator, organizer or sustainer. And their inability to find a grand unifying principle has given rise to a growing consensus that there isn't one. Is it any wonder that Vaclav Havel would pin our societal problems to the fact that Science has failed to connect us with the most intrinsic nature of reality and natural human experience. It has instead become the source of a broad and pervasive agnosticism producing disintegration and doubt rather than integration and meaning. What is becoming undeniably clear is that science must search for a source of inspiration higher than itself or it must perish.

Most scientists remain oblivious to the problem claiming that they need more physics and must find a way to push the barrier further back to that first picosecond after the big bang where these elusive laws might be found. But a few scientists have begun to call for constructive change. In his book, *The Creative Cosmos*,[32] the physicist, Ervin Laszlo writes of an approaching revolution in science on the order of the Copernican Revolution that will produce the greatest scientific breakthroughs since the 16th century. According to Laszlo, the next great paradigm shift in scientific theory must be a 'Cosmological' revolution in the classical sense that Cosmology, once the mother of all science, embraces the whole of reality and describes its order.

Another eminent physicist calling for change is Roger Penrose. In a 1999 interview associated with the film, '*The Flow of Time*', Roger Penrose, Faun Flynn and the moderator discuss the problem of explaining time in physics. The relevant comments by Penrose on this subject are included below:

> *Roger Penrose: 'I think there's always something paradoxical about the way we seem to perceive time to pass and the way physics describes time.*

Roger Penrose: So this means that in a sense, the present past and future are out there, and that also gives us a very deterministic view of the world. We have no control of what happens in the future because it's all laid out. I think the trouble that people have with this idea is that you think the future is under your control, to some degree, and so this means that if the future's laid out then in a sense it's not under your control.

Roger Penrose: The question of the passage of time is something the scientists have rather set aside, and taking the view that it's not really physics, it's a subjective issue; and subjective questions are not part of science. Now when you start talking about phenomena like one's own perception of the passage of time, then that is a subjective thing. And that's almost a taboo subject for science because it's subjective. The physical world, at least according to Relativity, is out there, and there is no flow of time, it's just there; whereas our feelings (we have this feeling of the passage of time) are intimately connected to our perceptions.

Roger Penrose: My view is that there is some large scale quantum activity going on in the brain. Physics does not say that Quantum Mechanics takes place in small areas, but also takes place over larger areas. I think this has to do with the consciousness. I think we need a new way to look at time, not either Quantum Mechanics or Relativity.

Roger Penrose: I don't think we have the tools, I don't think we have the physical picture to accommodate these things yet. We're not very close to it'.[33]

Interview concerning '*The Flow of Time*' (1999), Roger Penrose. [underlining emphasis mine]

Taken together, both Laszlo and Penrose envision the pressing need for a revolution in science, a new Cosmology that embraces the whole of reality and explains its order. Penrose writes of moving beyond the taboos of physics and dealing with things like subjectivity and consciousness. He even says that '*...we need a new way to look at time, not either Quantum Mechanics or Relativity*'. But he admits that we *'...don't have the tools...and we're not very close to it.'* Sadly, neither of these august scientists seems to understand that the tools they are seeking are right before their eyes in the laboratory of their own being. It is their mental instrument that cannot discover these elusive truths because, as Sri Aurobindo explained, '*...the mind seeks them as something other than itself and external to its own being.'*

Death of an Old Paradigm

I began this book by asking if science could explain the nature of our circumscribing reality and shed any light on its meaning or purpose. Clearly it cannot. Still '*...wandering in the satisfaction of the intellect and senses'*, science is lost, befuddled and cannot find its way. Even with the dawning knowledge that they have hit the wall, and can go no further in answering the important questions about our universe, scientists are unable to abandon their obsolete mental paradigm. They have become unwitting prisoners of a reductive mental model whose very nature is to reduce and fragment what spirituality assures us is inherently whole and interdependent.

> *'...Up until now, the philosophy of reductionism, which tells us how things work by breaking them up into parts has served us well. However the present crisis in particle physics can perhaps be expressed by the simple statement that reductionism and the search for grand unification seem no longer to be working.'*[34] **Lee Smolin, *The Life of the Cosmos*, Oxford University Press, 1997**

Given the fact that Science has become the driving force behind an increasingly secular society the consequences of its developmental paralysis are immense. Since the Mental paradigm is insufficient and inadequate to express reality in anything better than bi-polar oppositions, paradoxes and irreconcilables, a society which has mind as its highest instrument can end up in only one way, a world that destroys itself. Its reductive model leads inevitably to the destructive fragmentation of human nature and a civilization that has NO CENTER. One of the most prescient images of a world in this condition can be found in Yeats' iconic poem, '*The Second Coming*'.

> *'Turning and turning in the widening gyre*
> *The falcon cannot hear the falconer;*
> *Things fall apart; the centre cannot hold;*
> *Mere anarchy is loosed upon the world,*
> *The blood-dimmed tide is loosed, and everywhere*
> *The ceremony of innocence is drowned;*
> *The best lack all conviction, while the worst*
> *Are full of passionate intensity.*
>
> *Surely some revelation is at hand;*
> *Surely the Second Coming is at hand.*
> *The Second Coming! Hardly are those words out*
> *When a vast image out of Spiritus Mundi*
> *Troubles my sight: a waste of desert sand;*
> *A shape with lion body and the head of a man,*
> *A gaze blank and pitiless as the sun,*
> *Is moving its slow thighs, while all about it*
> *Wind shadows of the indignant desert birds.*
>
> *The darkness drops again but now I know*
> *That twenty centuries of stony sleep*
> *Were vexed to nightmare by a rocking cradle,*
> *And what rough beast, its hour come round at last,*
> *Slouches towards Bethlehem to be born?'*[35]

William Butler Yeats, The Second Coming, 1919.
[underlining emphasis mine]

Like all seers, Yeats understood the dynamics of chaos and could easily see the signs of a civilizational collapse brought about by the inherent contradictions within the mental consciousness. He was aware that Mind has only a limited range of effectiveness. When it reaches its vibrational ceiling so to speak, we begin to see a breakdown that manifests in three stages: Polarization, Contradiction and finally Paralysis. When we see these symptoms appearing in the world through events such as mounting political and social polarization, systemic economic collapse, violent religious fundamentalism, and escalating tensions between races and classes, it is clear that certain dangerous limits are being reached. Moreover, the growing passions of these oppositional camps and the almost equal division of their numbers testify that the process is quickly approaching a critical mass. These chaotic events are indisputable signs of an approaching mental apocalypse. Mind is being dislodged from its ruling position of many thousands of years and from all indications, these are its last days, perhaps its last hours. And without a radical change in the sphere of human consciousness the global catastrophe toward which we are rapidly moving will be unavoidable. The Mother explains:

> *'...One thing seems clear: humanity has reached such a generalized state of tension – tension in effort, tension in action, tension even in daily life – with such an excessive hyperactivity, such an overall restlessness, that the species as a whole seems to have reached a point where it must either burst through the resistance and surge forth into a new consciousness, or else sink back into an abysm of obscurity and inertia. This tension is so total and so generalized that obviously something must break. It cannot go on like this. Yet all this is a sure sign that a new principle of force, consciousness and power has been infused into matter and by its very pressure has produced this acute state. Outwardly, we might expect to see the old habitual means used by Nature*

> *whenever she wants to bring about an upheaval; but here there is a new phenomenon, which is evidently visible only in a select few…the will to find a new, a higher, an ascending solution, an effort to surge forth into a vaster, more encompassing perfection. Certain ideas of a more general, more extensive, more collective nature, as it were, are being worked out and are at work in the world. And the two go together: a greater and more total possibility of destruction and an inventiveness that unrestrainedly increases the possibility of catastrophe, a catastrophe that would be much more massive than it has ever been; and at the same time, the birth, or rather the manifestation, of much higher and more comprehensive ideas and wills which, when heard, will bring a vaster, more extensive, more complete and more perfect solution than before. This struggle, this conflict between the constructive forces of an ascending evolution, of an increasingly perfect and divine realization, and the more and more destructive forces – powerfully destructive, forces of an uncontrollable madness – is becoming more obvious, unmistakably visible, and it is a kind of race or battle as to which will be first to reach its goal…* '[36] **Collected Works of The *Mother*, First Edition, Volume 09, pp. 296-301[underlining emphasis mine]**

The *'ascending solution'* of which the Mother speaks is a New Way for humanity, one that provides an entirely new direction than all previous spiritual paths. The principle goal of this New Way is to restructure the consciousness away from its exclusive Mental/egoic orientation, and repoise the entire being, upon a higher Center; a **'Center That Holds'**.

The New Way

To appreciate the evolutionary necessity of a New Way we must listen carefully to the Mother's words. *'…all this [tension] is a sure sign that a new principle of force, consciousness and power has been infused into matter and by its very pressure has produced this acute state....'* But exactly what does she mean by 'infused'? Sri Aurobindo described it as the descent of a higher, 'Supramental' consciousness into the lower mental field, a new evolutionary force that is displacing mind in the process.

> *'...And when this higher power is coming down', said the Mother, 'it comes as an absolutely luminous and perfect organization which one can see if he has the vision...it presses on matter and everything begins to seethe and resist. It is the inertia and resistance which causes the catastrophe. It is not that the catastrophe was foreseen or caused, it is caused by resistance...It is one and the same movement of consciousness that expresses itself in a nature ridden with calamities and catastrophes and in a disharmonious humanity. The two things are not cause and effect but stand on the same level. Above them there is a consciousness seeking for manifestation and embodiment on earth, and in its descent toward matter, it meets everywhere the same resistance in man and in physical nature. All the disorder and disharmony we see upon the earth is the result of this resistance. Calamity and catastrophe, conflict and violence, obscurity and ignorance – all come from the same force.'*[37]**The Mother - *Questions and Answers*, May 5, 1929, *Collected Works of The Mother*, Sri Aurobindo Ashram Press, India,[underlining emphasis mine]**

Clearly the greatest sin in this new world of the Supermind is *'tamas'* - inertia, ignorance, obstruction and mental fundamentalism of all stripes. Yet we see this resistance everywhere; in the dying grasp of dictators, the stubborn greed of out of control capitalists, the fanatical fury of religious zealots, and science maddened elites seeking to impose their technology on every single aspect of human existence. But who would have imagined that the upsurge in climatic and natural disasters throughout the world was a consequence of this same descending force. The implications of what the Mother has written are truly terrifying when you understand the extent and rigidity of the resistance we face. But at the same time there are emerging forces for change. The 'Arab Spring' and 'Occupy' movements are living examples of this budding consciousness beginning to display signs of its future potential, but it is still in its nascent stages and not yet grounded in a coherent evolutionary vision. In order for the individual to become an effective and conscious instrument of this Force, he must undergo an extensive transformation and learn to SEE in a completely new way. Failing that, his efforts, no matter how well intended are subject to deviation by the mental ego and will not produce the desired results. To prepare the individual for a new poise of consciousness through which the supramental power and force may flow unimpeded a process of yoga is required. But this cannot be done through the old familiar paths. It is only by immersion in the cosmology and Knowledge unveiled throughout Ms. Norelli-Bachelet's New Way that we can forge the necessary alignments which give rise to a 'new seeing'.

If we consider the nature of human awareness as a hierarchical phenomenon ranging from the ordinary egoic consciousness to a supreme enlightenment or gnosis, there must be certain modes of perception that define that spectrum. Indeed, each of the gradations of consciousness sees the world through a certain lens or filter and nothing is allowed to enter our consciousness that does not reinforce that particular view of reality. This is clearly evident in the empirical prejudice of the scientific observer. As we move up the spectrum however, we begin to dispense with our filters and come to see the world as it truly is.

For most people this process of refinement is not clearly understood and can take many lifetimes to achieve. But there is a knowledge by which one can undertake a faster more deliberate evolution. The very existence of this knowledge has been the stuff of legend for thousands of years because it was believed to contain the mystery that could turn lead into pure gold. The 'Philosopher's Stone' as it was called contained the most coveted secrets of the age. Its attainment was the supreme goal of knowledge: to perceive and know with a certainty that one knows; and to see with an understanding which can be infinitely extended to all situations. In the same way the Vedic mysteries had been preserved for the proper time of their unveiling, the alchemical secrets of the Philosopher's Stone were also preserved until they could be openly revealed to humanity.

The Philosopher's Stone

As we move beyond the human consciousness and toward the Supramental there are three fundamental realizations that are indispensable for a successful journey. If understood, they are the means of a radical shift in consciousness; if not, they represent a kind of natural boundary beyond which we cannot go. These three have to do with the Structure of the Universe, the Unity of all existence, and a New Model of Time. All three are combined in Ms. Norelli-Bachelet's New Way which deals with what, for centuries, has been divorced from traditional spirituality - the question of Time and Measure. The centerpiece of her work is a cosmological model which highlights certain essential relationships between man and the universe he inhabits and provides the means by which one can unveil a common center.

We touched upon the first of these realizations in Ms. Norelli-Bachelet's discussion of the nature of an Axis in which she mentioned a system of ordered correspondences ranging from the Cosmic, Universal, or Solar systems down to a nation, race, or individual. In this brief statement she introduced us to one of the most important truths of creation known as the Law of Correspondence and Equivalence. The correspondences which exist between the Individual and the Cosmos have been known for millennia but in this age of the Supermind, the nature of these correspondences have been updated to a level of precision and clarity unknown in ancient times.

One of the earliest accounts of this hidden knowledge was the famous Emerald Tablet (*Tabula Smaragdina)*, attributed to Hermes Trismegistus. It was an ancient alchemical text purporting to reveal the deepest secrets of creation. The Tablet became the core of an alchemical teaching that influenced a long line of spiritual seekers including Aristotle, the Rosicrucians and Sir Isaac Newton. Among its many secrets was the principle of correspondences expressed in a simple yet utterly profound formula:

> *'That which is Below corresponds to that which is Above, and that which is Above corresponds to that which is Below, in the accomplishment of the Miracle of One Thing. And just as all things have come from One, through the Mediation of One, so all things follow from this One Thing in the same way.'* [38] **Hermes Trismegistus**

The physicist David Bohm brought this ancient concept into the modern era with his theory of a holographic universe. Bohm concluded that if one is able to remove his usual mental lenses, he will be able to see the universe as a hologram in which each part contains all the patterns possessed by the whole. This was celebrated as a brilliant new concept but in reality, it was anything but new. It had been adopted thousands of years ago as a foundational principle of Vedic cosmology and was communicated through a symbolic image known as the *'Net of Indra'*. As one of the chief deities of the Rig Veda, Indra was known as the Lord of Heaven. Over his palace on Mount Meru hung a jeweled net which had a multifaceted gem at each vertex. Each jewel was reflected in all of the other jewels. Everything that existed in Indra's net implied all else that exists. It was an image of pure holographic unity. In his book, '*Yoga, Immortality and Freedom*', the renowned scholar of world religions, Mircea Eliade wrote about these principles of correspondence and put them forward as a means to mystical unity.

> *'...We must never lose sight of the fact that the tantric universe is made up of an endless series of analogies, homologies, and symmetries; starting from any level, one can establish mystical communication with the others, in order finally to reduce them to unity and master them.'*[39]
> ***' Mircea Eliade,Yoga, Immortality and Freedom***

What can be appreciated in these images and quotes is an attempt by spiritualists both ancient and modern to find the key to 'Grand Unification', a theory that might at last explain the workings of the universe. Their quest parallels what modern physics has been seeking for decades, albeit unsuccessfully. What eludes them both is the perfect jewel of the human Soul wherein the code of creation is written. But in order to make that discovery we need to look more deeply into the creation itself and the structure of our manifest universe.

The Axis and the Plane

Everything in the observable universe, Galaxies, Solar systems, Planets, Individuals, and even Elementary Particles, rotate on an **Axis**. And everything that rotates on an axis brings into being a center that holds a field or **Plane** that extends out from that axis to form an ordered Cosmos. We can see this principle in our own Solar system. The Sun at the Center rotates on its axis which holds in place an orbital plane that extends out from itself to the far edges of the solar system. All of the planetary bodies were formed in that plane and continue to orbit the Sun in an unending elliptical harmony. Within that celestial harmony the Earth rotates on its own axis carrying the Moon in its field as it orbits around the Sun. Our own solar system is located on the horizontal plane/disk that extends out from the galactic center of the Milky Way. And in the same way that our solar system orbits that greater center, the Milky Way is in orbit with 50 or so other galaxies around a massively dense galactic cluster called the 'Great Attractor'. And who knows how many centers orbiting greater centers there are before we get to that original Singularity from which everything emerged. If we stop and think about it we will realize that 13.7 billion years ago the entire universe emerged from a point/Axis/singularity and began to spin. Out of that primordial chaos order began to emerge as a result of that spinning. Super Galaxies began to form and rotate around that original Singularity. Galactic clusters began to form and rotate around the Super galaxies. Galaxies began to form and rotate around those clusters and Solar Systems began to organize themselves within the galactic planes. And from there planetary bodies like our Earth emerged and began to spin on their axes around their local Suns. If we apply the axiom 'as above, so below', everything we can see - from the greatest galaxy down to the most minute elementary particles - is formed from One pattern, an original template creating self-similar forms on all scales like the eight Russian Matryoshka dolls, nesting one within the other. We will find this same fractal image in the famous vision of the biblical prophet Ezekiel:

> *'...And I looked and behold a whirlwind came out of the north. A great cloud and a fire and out of the midst thereof came the likeness of four living creatures. And every one had four faces and every one had four wings. As for the likeness of their faces, they four had the face of a Man, and the face of a Lion, on the right side, and they four had the face of an Ox on the left side and they four also had the face of an Eagle. Their appearance and their work was as if a wheel within a wheel; as for their rings, they were so high they were dreadful and their rings were full of eyes, round about them four. And when the living creatures went, the wheels went with them for the spirit of the living creature was in the wheels.'*[40] ***The Holy Bible*, Ezekiel: Chapter 1. [underlining emphasis mine]**

In the Vedic scriptures written thousands of years earlier, you will find the exact same language describing the process by which *Vishnu Trivikrama* measures out the universe. Vishnu makes three great strides which, including his starting point, cover each quarter of the great circle.

> *'1. Of Vishnu now I declare the mighty works, who has measured out the earthly worlds and that higher seat of our self-accomplishing he supports, he the wide-moving, in the threefold steps of his universal movement.*
> *2. That Vishnu affirms on high by his mightiness and he is like a terrible Lion that ranges in the difficult places, yea, his lair is on the mountain-tops, he in whose three wide movements all the worlds find their dwelling-place.*
> *3. Let our strength and our thought go forward to Vishnu the all-pervading, the wide-moving Bull whose dwelling-place is on the mountain, he who being One has measured all this long and far-extending seat of our self-accomplishing by only three of his strides.*
> *4. He whose three steps are full of the honey-wine and they perish not but have ecstasy by the self-harmony of their nature; yea, he being One holds the triple principle and earth and heaven also, even all the worlds.*

5. May I attain to and enjoy that goal of his movement, the Delight, where souls that seek the godhead have the rapture; for there in that highest step of the wide-moving Vishnu is that Friend of men who is the fount of the sweetness.
6. Those are the dwellings-places of ye twain which we desire as the goal of our journey, where the many-horned herds of Light go traveling; the highest step of wide-moving Vishnu shines down on us here in its manifold vastness.' [41]***RigVeda 1.154*** **[underlining emphasis mine]**

To further illustrate the universality of this Cosmic vision of 'wheels within wheels' and the classic zodiacal images dividing those wheels into four parts we need only read the Revelation of Saint John the Divine written on the Greek island Patmos, around 70 AD.

'I looked and behold, a door opened in Heaven; and the voice said, come up hither, and I will show you the things which must be hereafter. And immediately I was in the spirit: And beheld a throne was set in Heaven and One sat on the throne. And round about the throne were four and twenty seats and upon the seats were four and twenty elders clothed in white raiment and on their heads, crowns of gold. And in the midst of the throne were four beasts with eyes before and behind. And the first beast was like a Lion [Leo] and the second beast like a calf [Taurus], and a third beast had a face as a man [Aquarius] and the fourth beast was like a flying eagle [Scorpio]. And the four beasts had each of them, six wings about him and they were full of eyes within: and they rest not day and night, saying holy, holy, holy, Lord God Almighty, which was and is and is to come.' [42] ***The Revelation of Saint John the Divine, Chap. 4; 1-9.*** **[underlining emphasis & brackets mine]**

'With respect to that same cosmic sea, the visionary sees four `beasts' therein: the first is a Lion, the second is a Calf, the third is a Man, and the fourth an Eagle. If the

Eagle, the fourth sign, was left out of Vishnu's measuring it is because this Eagle is Garuda, his own carrier. He begins his measuring from that point in the wheel, also known as Scorpio, and takes `three steps'. Scorpio, otherwise known as the zodiacal Eagle, would be the fourth in correct sequence, similar to John's text.' [43]
Patrizia Norelli-Bachelet, *The Hidden Manna*, Aeon Books, 1976

According to Ms. Norelli-Bachelet, the entire mystery of creation is contained in the knowledge of these 'wheels within wheels' that spin around a common axis. But in order to fully understand this omnipresent principle of an Axis and Plane we must first discover from whence it came. To date, neither Physics nor Metaphysics has been able to shed any light on this question. Physics cannot explain the nature of an axis because there is nothing there to measure. There is only its function, a subtle 'pole' that holds everything together. And metaphysicians, bent on dissolving themselves in a transcendental absolute, are in total denial of the existence of any soul/center/axis. For them the question does not even arise. Fortunately for us, the Vedic Seers did not experience those kinds of limitations.

In the Rig Veda Mandala 1, Hymn 164 the Rishi reveals something of this wonderful mystery. Speaking of the Earth's axis, he writes: '*…The subtle axle of the earth does not get rusted and the earth continues to revolve on its axle.'* Another translation says: *'...Its axle, heavy-laden, is not heated: the nave from ancient time remains unbroken.'*[44] In this passage we are given precious information. The axle is eternal, subtle, it does not rust. It is heavy laden yet frictionless and is not heated. The axle operates in conjunction with the *'nave from ancient time'* which translates as the central block or hub of a wheel. It is clear that the seers are writing about realities that are quite invisible to the material eye. They were not made from empirical observation as science has done. Everything was discovered internally. They found this immutable Axis within themselves and sang its praises calling it the *'Fulcrum of Creation'*, *'Skambha'* - the *'cosmic pillar'* upholding the worlds. And most importantly, they declared in Mandala 10, Hymn 149 that the entire universe was born and developed from a central point…the Hub of the Vedic Wheel.

Twelve spokes, one wheel, navels three.
Who can comprehend this?
On it are placed together
Three hundred and sixty like pegs.
They shake not in the least. [45]
(Rig Veda 1.154.48)

The Vedic Wheel

One is the wheel; the bands are twelve;
three are the hubs – who can understand it?
Three hundred spokes and sixty in addition
have been hammered therein and firmly riveted...

<u>*Though manifested, it is yet hidden, secret,*</u>
its name is the Ancient, a mighty mode of being;
in Skambha is established this whole world;
therein is set fast all that moves and breathes. [46]
(Atharva Veda 10. 8)[underlining emphasis mine]

In these curious passages the Rishi is describing an Axle and a Wheel. 'Who can comprehend this?' he asks. For thousands of years this question has remained unanswered because it deals with the most profound and invisible realities of all time. Light can only be brought to this subject by a seer of an extraordinary order, one who has pierced into that luminous seed that holds the secrets of the Supermind. Ms. Norelli-Bachelet is the only person to unveil these mysteries since the Vedic Age. She explains to us that what the Rishi has 'seen and heard' on the plane of truth is nothing less than a complete revelation of the Axis and the Plane, the fundamental truth-components of our existence.

"The Axis", says Ms. Norelli-Bachelet, *"is our link to the 'other side'; it is the 'center that holds' that gives rise to the Plane. The Plane could not come into being without the Axis, and the Axis has no reason for existence without the Plane."* [47] The Axis represents the Contracting force of Time, Unity, and Being, while the Plane represents the expanding power of Space, Multiplicity and the Becoming of that same Being.

To better understand the nature of the Plane we must return our thoughts to the Supramental formula: 9-6-3-0/1 which describes the procedure by which solidified energy, form and matter emerge from the Singularity to begin its orderly evolution in space as our material universe. This can be easily understood through the metaphor of human childbirth. Like a mother in the throes of labor, the action in the zero womb is one of intense contraction which compacts the triadic forces 9, 6, and 3 - the past, present and future, and orders them in such a way that they could only emerge into manifestation as a unity of self-similar geometric structures of consciousness simultaneously expressing itself as the Transcendent, Cosmic and Individual. Once freed from the crushing contractions, just as an infant experiences when out of the mother's womb, the One is born and extends itself in Space through the impulsive power of expanding Time. The occult order of things compacted in that zero womb once it crosses the event horizon of Space can be seen in the structure of the Vedic wheel, which perfectly describes the horizontal Plane. The geometry of its design captures the flow of Time in the field of Space.

*'One is the wheel '*says the Rishi; *'...the bands are twelve; three are the hubs ... three hundred spokes and sixty in addition have been hammered therein and firmly riveted...Who can understand it?'* For those with eyes to see, the central point of the wheel represents the Absolute upholding the play of creation. It can also be the 5th element known as the Etheric. The three concentric circles or 'hubs' are the three energy flows or 'Gunas': Creation, Preservation and Dissolution. The four inner petals represent the four 'Tattvas' or elements: Fire, Earth, Air, and Water, each of which is expressed in three energetic modes. The four elements multiplied by the three energy flows or gunas give rise to the twelve outer petals which from antiquity have been known as the 12 signs of the Zodiac and which are located on the outer band of the 360 degree circle. To the seer, the Axis and Plane form a coexistent and seamless vision of Spirit becoming the material expression of itself. And it does this in a way that allows us to follow the evolving in our Time measure of what was involved in that zero womb by compression or contraction.

Churning of the Milky Ocean at the Dawn of Time

In the Vedic period the process of differentiation and densification of consciousness into material forms was explained in the form of a simple myth. The static form of the Axle and Wheel is made dynamic in a tale known as the *'Churning of the Milky Ocean at the Dawn of Time'*[48]. It recounts the same process by which the chaotic primordial soup of creation became ordered as galactic, solar and human systems. The organizing principles of the Churning myth may also be applied to a world in upheaval similar to our own - a society on the border between a fragmentary and unitary system in which the natural order has begun to degenerate producing such a generalized state of tension and chaos that something has to change. If we follow the lines of this extraordinary tale it will provide us with a model of our own evolutionary crisis as well as a complete cast of characters. The delicate balance of opposing forces in the myth is symbolized by the gods in their struggle against the demons. Good and evil, light and darkness, and creation and destruction are all part of this cosmic process and are included in the play. When order begins to collapse, the Demons, ever watchful for an opportunity to gain the upper hand, seize the moment and chaos reigns. Destructive forces of an uncontrollable madness take the day and things of great value are lost. The very existence of the universe is threatened. But after complaints from Indra, the warrior king of heaven, to Brahma, the creator, the god Vishnu is called in to restore the balance and the cosmic order.

The Churning of the Sea Milk, painting, Punjab Hills, 19th Century

According to the myth Vishnu, the preserver, devised a clever plan to restore order. His idea was to activate the dynamic equilibrium by churning the great Milk Ocean using Mount Meru as a churning stick or Axis. However, the weight of Mount Meru was such that it began to sink into the soft bed of the Milk Ocean and so Vishnu assumed the form of the Tortoise, Kurma, whose curved back became the stable support and pivot upon which the churning stick could rest. He then called on Vasuki, the primordial serpent of Time to wrap himself around the Mount Meru axis, as a churning rope. Following a ruse by Vishnu which convinced the demons that the gods wanted to hold Vasuki's head, they became irate and insisted that the demons should take the head and the gods the tail. Thus the demons and the gods took opposite ends of the cosmic serpent and the churning process began. Unbeknownst to the demons, as the churning progressed Vasuki's breath became hot and out of his mouth came poisonous fumes which suffocated the demons while the gods, at the tail, were refreshed by cool ocean winds. As the churning of the Milk Ocean continued, treasures began to appear as butter might emerge from the churning of cream. Most important of the treasures was Amrita, the nectar of immortality. Kamadhenu, the Vedic Cow of Plenty also emerged from the churning process as did Airavata, a beautiful white elephant and Uchchaisravas the white horse. The four Vedas also arose from churning the Milk Ocean. However, as these treasures arose so did a poison which the myth tells us was consumed by Shiva in order to save the world. The poison was caught in Shiva's throat which turned his neck blue but it also purified the serpent Vasuki which Shiva thereafter wore as his girdle.

This cosmogonic myth contains the details of a profound occult process. It illustrates the 'alchemical' dynamic between two opposing forces which generates the passage from chaos to cosmos, order out of disorder. The most important aspect of the churning tale is the creation of an Axial alignment, a stable point around which the dynamic motion occurs.

> *'...The concept of the "Axis" is the most important cosmic realisation ever to come into being and it lies at the core of the Rishis' obsession with alignment. In fact, one could say that all of Vedic spirituality is an expression of this axial alignment.'*[49] **Patrizia Norelli-Bachelet, Letter to RamachandraRao.**

The elements of this tale are extraordinarily simple: the churning stick is the vertical Axis, the Milk Ocean is the horizontal Plane, and Time is the cosmic serpent Vasuki who sets the process in motion. As the churning progresses treasures emerge like butter rising from the working of the cream. Consciousness densifies and gifts appear which can only come into being through the dynamism of the creative process. And out of the churning and all that follows comes a new cosmic alignment. Order is once again restored and harmony rules throughout the worlds.

An Applied Cosmology

Like the Solar and Planetary systems, the Individual also possesses an axis and a plane but owing to his off-center poise, they exist *in potentia*, not as an active conscious feature of his experience. This is due to the fact that up until now the structure of human consciousness has been founded on the Mental principle. And since the mind is insufficient to see and understand the true nature of reality it organizes itself around an inferior poise of consciousness known as the Ego. It is exactly as Plato explained, the ego mind is limited to the mere appearance of things, shadows of a greater reality. And because the ego cannot see, it craves for lesser vital satisfactions (usually related to power, money and sex) of which it can never get enough. And this is the tragedy of our present mental species, a collectivity made up of shadow creatures whose lives orbit a meaningless void from which they seldom escape. In his poem, *Savitri*, Sri Aurobindo laments the mental human condition:

'O Force-compelled, Fate-driven earth-born race,
O petty adventurers in an infinite world
And prisoners of a dwarf humanity,
How long will you tread the circling tracks of mind
Around your little self and petty things?
But not for a changeless littleness were you meant,
Not for vain repetition were you built;
Out of the Immortal's substance you were made;
Your actions can be swift revealing steps,
Your life a changeful mould for growing gods.

A Seer, a strong Creator, is within,
The immaculate Grandeur broods upon your days,
Almighty powers are shut in Nature's cells.
A greater destiny waits you in your front:
This transient earthly being if he wills
Can fit his acts to a transcendent scheme.
He who now stares at the world with ignorant eyes
Hardly from the Inconscient's night aroused,
That look at images and not at Truth,
Can fill those orbs with an immortal's sight.
Yet shall the godhead grow within your hearts,
You shall awake into the spirit's air
And feel the breaking walls of mortal mind
And hear the message which left life's heart dumb
And look through Nature with sun-gazing lids
And blow your conch-shells at the Eternal's gate.
Authors of earth's high change, to you it is given
To cross the dangerous spaces of the soul
And touch the mighty Mother stark awake
And meet the Omnipotent in this house of flesh
And make of life the million-bodied One.
The earth you tread is a border screened from heaven;
The life you lead conceals the light you are.'[50]
Sri Aurobindo, *Savitri - A Legend and a Symbol*, page 370
[underlining emphasis mine]

In this exquisite passage, Sri Aurobindo reveals the radiant evolutionary possibilities that lie beyond the imprisoned ego. Moreover he gives us a breathtaking clue to the process by which we might prepare ourselves for the glorious metamorphosis from human to the divine. The poet advises, '*...This transient earthly being, if he wills, can fit his acts to a transcendent scheme.*' Here the earthly traveler is given a choice, a new life and a new seeing, but one depending on his willingness to adjust his actions so that they conform to a higher model of consciousness. You could not find a more succinct description of Ms. Norelli-Bachelet's New Way.

Oneness

We have discussed the first element of this *'transcendent scheme'* which is a thoughtful embrace of the principles of correspondence and equivalence that reveal the true structure of the universe. The second element of the New Way which follows upon the first is a willingness to seek and discover the hidden Unity of All Existence. Sri Aurobindo provides us with the nature and means of our goal:

> *'...The express direction of the path of Yoga is the highest unity, the call to absolute Oneness. The self of the individual must be made one with the Self of all. This cannot be done without an uncompromising abolition of the Ego-sense at its very basis and source. One attempts this abolition, negatively by a denial of the reality of the ego, and positively by a constant fixing of the thought upon the idea of the One and the Infinite everywhere. This, if persistently done changes in the end the mental outlook on oneself and the whole world and there is a kind of mental realisation; but afterwards and by degrees the mental realization deepens into spiritual experience - a realisation in the very substance of our being.... '* [51]**Sri Aurobindo, *The Synthesis of Yoga*,p. 362.**

The only way for absolute Oneness to become an unshakeable fact of our existence is through the direct lived experience. And this is best accomplished through an intelligent and progressive process of 'Surrender'. But this can be extraordinarily difficult for most people because the mental ego considers itself the doer and decider of our lives. And when the ego is at the center of our experience, it creates certain ways of understanding, feeling, and looking at life that obstruct a perception of oneness. As if by enchantment, the ego sees the false as true and nothing is allowed to enter our consciousness that does not reinforce that illusory view. When this misaligned perspective is shared at a collective level, as it is in our contemporary culture, enormous pressures are brought to bear to make sure everyone conforms to these deceptive ego values. And these expectations become the very chains that imprison the lost souls of Plato's underworld.

The problem for the individual is the same for science - neither can conceive of anything higher than itself to bring meaning, purpose and order to their experience. The idea of surrendering control of your life to something you cannot see and measure is viewed by the secular establishment as shear madness or worse. But the intrepid seeker of the New Way understands what he is surrendering to and appreciates the need to courageously risk himself again and again until that which is indestructible can arise within him. This may require abandoning all plans and security and going forward without knowing what the future will bring. It means standing in the midst of the world of opposites and refusing to cling to any external supports. If one can endure this churning process with patience, perseverance and absolute faith, a new alignment or 'Axis' comes into being. Once this new center is forged one makes the astonishing discovery that one has become aligned with the infinite intelligence that governs all of nature's movements, a consciousness of Oneness which knows itself each instant everywhere. In a 'seeing' of this nature there also comes the understanding that all is arranged, all is controlled, and all is a perfection. When this vision of Oneness becomes a settled thing in our being, we are ready for the real and true transformation to occur.

The Supramental Time Vision

The first two realizations of the New Way, the truths of Correspondences and Oneness, represent mysteries which have been known and followed by seekers for millennia. And in most cases their accomplishment, along with lesser mysteries of the old religions, marked the end of the spiritual quest. But this is not, as many believe, the end of man's evolutionary potential with nothing beyond to explore. It is the end of the lower path and the beginning of the higher evolution. In the New Way, the truths of Correspondences and Oneness are necessary prerequisites for the third element that distinguishes it from all previous paths: a New Vision of Time. Over the ages seekers have fled from this vision of the Time-Spirit seeking instead the static peace of transcendence, a nirvanic extinction, or the promise of an after-death heaven. And the collective pursuit of these strategies of escapism has left an unbridgeable chasm between the spiritual and physical worlds. Just as night follows day, this evolutionary defect has led to overwhelming numbers of people living out a meaningless, purposeless, and obsolete existence. In order to move beyond this developmental cul-de-sac it became necessary to find a New Way to confront the most stubborn impediment of the human ego - its illusory perception of Linear Time. The solution to this age old problem is Ms. Norelli-Bachelet's gift to the world.

Due to his off-center poise, the average individual whose mind has evolved in time sees nothing unique about the temporal order. He becomes unconsciously entwined with Time's apparent linear character of past, present and future and does not realize that his life unfolds in a succession of archetypal cycles. As Ms. Norelli-Bachelet observes, *'...We live in the midst of something we completely ignore. Our eyes are veiled and our bodies serve to secure that the veils remain intact.'* But by working with time in a particular manner, it is possible to restructure the consciousness, and ultimately the physical being, around a higher axis. This was one of the greatest mysteries of the Rig Veda but one that remained unrecognized because for thousands of years seekers were encouraged to deny and reject time in order to lose themselves in a Timeless Absolute. It was not until the appearance of Sri Aurobindo and his line that the true nature of Time could be revealed.

> '*...It cannot be denied that Time is a major feature of our material universe. In consequence, it stands to reason that when spirituality decides that the universal manifestation is unreal and illusory, Time is immediately shunned and the seeker is required to go 'beyond time', i.e., to the experience that appears to obliterate Time. Conversely, if we are engaged in a path that seeks a transformation of matter and an experience of the highest planes of reality in this material creation we are obliged to accept the instrumentation of Time. Furthermore, we must seek a means or a key by which Time itself becomes the element utilised in the gestation of a new and superior consciousness-form in the material creation. Hence a new harmony of time results, the essence of which is contained in the [Supramental Cosmology].*'[52] **Patrizia Norelli-Bachelet, *The New Way, Vol. 3*, 1983, [underlining emphasis mine]**

Just as the material universe unfolds as wheels within wheels, Time unfolds as cycles within cycles. It has a fixed course and a given order and extends itself in a series of cycles repeating themselves on many different levels of experience. The means by which Time becomes an ally in the gestation of a new consciousness becomes apparent when the individual aligns himself with these cycles and become progressively imprinted with their rhythms and harmonies. It sounds so simple yet according to the Rishi, this is the secret of secrets:

> *'Certain eternal worlds are these which have come into being, their doors are shut to you (or opened) by the months and the years. Without effort one world moves in the other, and it is these that Brihaspati has made manifest to knowledge.'* [53] ***RigVeda II.24.5***

The initiatic journey of the Rishi is a journey in Time through the Year and its Twelve Months, the central figure of the Veda around which the yogic Sacrifice is conducted. It was proclaimed as the greatest mystery which opened the doors to the Supreme light. But these eternal worlds have been closed to us, just as they have been closed to science, by our misperception of the true nature of time. To reverse this crippling error and recover a vision of the sacred, the months and the years have to be re-discovered and created in us by that very same power of Time. This is most easily achieved through the application of a cosmological model which illumines certain essential relationships between man and the universe he inhabits. The most important of these relationships lies in understanding the sacred rhythms of the Earth as she moves in her never-ending orbit around the Sun.

The Cycles of the Earth

The Earth, our home and mother, is 4 ½ billion years old. It is the third planet in our solar system and like the other planets it orbits the central Sun in an eternal elliptical path. The Earth moves from west-to-east around the Sun following a path known as the 'Ecliptic plane'- so named because the eclipses occur only when the Moon is on or near this circuit. The Earth rotates on an invisible axis which is tilted at 23 ½ degrees measured from the perpendicular to the Ecliptic and it is this tilt that causes us to experience the Equinoxes and Solstices. These in turn produce the Seasons, each consisting of three months, a period that corresponds to the energy flows or 'gunas'. The axis thus defines the **four cardinal points** that divide the Earth's circuit in the flow of the year in its revolution around the Sun. The Cardinal points are the Equinoxes of equal days and nights, and Solstices of unequal measures. Together they provide the structure for the experience of the four seasons on Earth. This structure is permanent for our planet: it is eternal and unchanging.

According to astronomical observation, the Earth has five significant movements. It rotates on its axis at 1,038 mph to produce a daily rotation of 24 hours. It orbits the Sun at a rate of 66,500 mph making a complete revolution every 365.25 days. Owing to the tilt of its axis the Earth also has a Seasonal cycle with Spring occurring around March 21 on the Vernal Equinox. The Autumnal Equinox occurs around September 21. And when the Sun is at its greatest distance from the equator (23.5 degrees) we experience the Solstices on which we observe the longest and shortest days of the year. The Summer Solstice occurs around June 21 and the Winter Solstice falls near December 21. In addition to these apparent movements, the Earth experiences a 41,000 year cycle during which its angle of tilt relative to the pole of the ecliptic moves from 22.1° to 24.5°. This cycle is believed to influence the long term heating and cooling cycles of the Earth that create the ice ages and the great floods when the build-up of ice subsequently melts. And finally, owing to a bulge at the Equator, the Earth's spin has a slight wobble that traces a hidden, invisible circle in the heavens that takes 25,920 years to complete. This great cycle is known as the 'Precession of the Equinox' and the Earth's movement by precession is what determines the astrological Ages, such as our present age of Aquarius. The discovery of the Precession was attributed to the Greek astronomer Hipparchus in the 2nd century BCE but it is clearly evident from the way that Vishnu measured the universe as well as other scriptures that this concept was already known and perfected thousands of years earlier in Vedic times:

> *'...Your two wheels, Surya [the Sun], the Brahmins know in their measured rounds. But the one wheel that is hidden, only the inspired know that.'* [54] **Rig Veda 10.85.16**

The reason it is so important to understand the Earth and its movements is because of the direct correspondences that exist between the cycles of the Earth and the life of the Individual. We have noted that the Earth's orbit around the Sun is divided into four quarters owing to the tilt of its axis, which causes us to experience the Solstices and Equinoxes - four points which mark off the stages of the great cycle of life. Spring, Summer, Fall and Winter are the seasons that organize all life on Earth from birth to death to rebirth in an eternal cycle. They represent the ebb and flow of energy-consciousness in various stages of manifestation. But it is not just the Earth that experiences orderly and purposeful seasons in its cycle, the individual follows an identical plan in the cyclic unfolding of his own life. Once again we are given correspondences that bring to light the structure and process of our own becoming. When we explore these symmetries we discover that the four seasons find an equivalence in the individual as he cycles through the four planes of consciousness and parts of his own being.

Twashtri's Bowl

In the Rig Veda there is a deity known as 'Twashtri' or Vishwakarma who is considered to be the architect of the gods and the shaper of forms. He is the artist and craftsman who gave structure to the perfect correspondences between man and the cosmos. Twashtri, it is said, fashioned a 'bowl' which was equal in structure to the celestial sphere and the planes of human consciousness. He declared this bowl to be the base of man's development from which he could ascend to the supreme heights. This bowl, while one, like the Earth's yearly rotation, was made into four, like the seasons of man. Twashtri shaped four bowls out of the original one, representing the four-fold planes of human consciousness: the Physical, the Vital, the Mental and the Spiritual/Supramental.

> *'...And this bowl of Twashtri, new and perfected, you made again into four.'*[55] **Rig Veda, 1.20.6**

> *'...For Twashtri, the Framer of things, has given man originally only a single bowl, the physical consciousness, the physical body in which to offer the delight of existence to the gods. The Ribhus, powers of luminous knowledge, take it as renewed and perfected by Twashtri's later workings and build up in him from the material of the four planes three other bodies, vital, mental and the causal or ideal body.'*[56] **Sri Aurobindo, *The Secret of the Veda*, p. 341**

Twashtri's bowl is just one of the mythic images that reflect the four inner petals of the Vedic wheel. This natural division of four is carried over into many aspects of the Vedic culture: the four Cardinal directions, the four Vedas, the four Yugas or Ages, the four aspects of the Divine Mother, the four Oceans in the Rig Veda and Atharva Veda representing the divisions of the celestial sphere, and the four Castes that define the order and division of human activity. As we move outward from the four inner petals, we see the circle extended and each quarter more specifically defined by a multiple of three (the three gunas or energy flows) which give us the twelve outer petals of the Vedic wheel. In the Veda this image of twelve petals becomes embodied through the myth of Aditi and the Ādityas. Aditi is the only goddess mentioned by name in the Rig Veda. She is the mother of the gods and the personification of universal all-embracing Nature. Aditi has twelve sons and daughters known as the Ādityas. Together they form the structure and signs of that universal image we know as the Zodiac.

> *'...Today we speak of the signs of the zodiac and the months they govern or represent; to this we add the science of number power, the calendar and the various systems of measuring space. We learn from this discussion, however, that these designations are no different from the terms used by the Vedic Rishis and that they too spoke of this one truth. The gods, children of Aditi, were rays of the twelve-rayed Sun and Aditi was this totality. Within those basic twelve rays there are nine planets, and these too are personified in the myths in the form of gods or helpers of the gods. Again, in the twelve original rays there are three different energy flows, which are also personified, - for example by the Ribhus. These latter, whom Sri Aurobindo describes as, 'the artisans of Immortality', have been called the 'rays of the Sun'. They are three in number and clearly connected with movement in the Rig Vedic hymns dedicated to them, for they fashion the steeds of Indra and the car of the Ashwins, 'And this bowl of Twashtri new and perfected you made into four. So establish for us the thrice-seven ecstasies, each separately by perfect expressing of them' (RV I.20.6,7) These three synonyms of movement are the energy flows that in the zodiac are repeated four times; hence it is said 'this bowl of Twashtri... you made again into four'. The thrice-seven ecstasies, or twenty-one ecstasies are the combined scales of 9 and 12...'*[57]

PatriziaNorelli-Bachelet, *The New Way*, pp. 401-2. [underlining emphasis mine]

The archetypal circle of the zodiac is an irreducible truth found in all cultures and all times. The myths, mysteries and monuments of all ancient civilizations were based upon this same Cosmic vision. But nowhere will you find a more precise elucidation of the zodiacal structure and principles than in the Vedic myths. In the passage above Ms. Norelli-Bachelet has given us the mythic symbols of the twelve 'rays' or divisions of the circle of Space. We can find this same design in the 12 Gods and Goddesses of the Greek and Roman pantheons. We can even see it in the figure of Christ and the twelve Apostles. But only in the Vedic pantheon will you find such precise personifications of the principles of Movement and Time. Twashtri, the Ribhus and the Ashwins are all given in association with another circle, the circle of 9 based on the 9 planets. Thus we have two circles, one of Space and the other of Time.

The Circle of Space

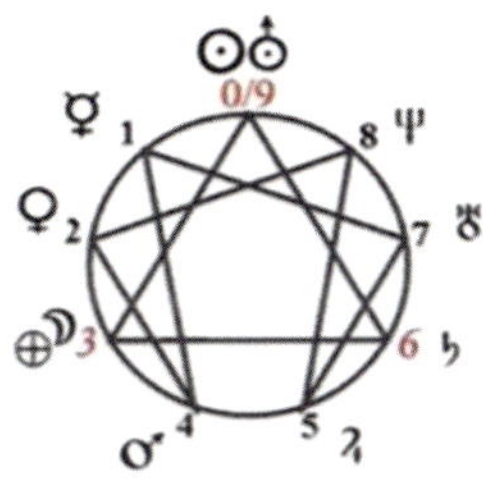

The Circle of Time

The Zodiac is the Circle of SPACE representing the Horizontal plane and expansion. Its circle of 12 is produced as a multiple of the 3 Gunas: Cardinal, Fixed and Mutable, and the 4 elements: Earth, Air, Fire and Water. This symbol contains the archetypal knowledge of the evolution. The *'Enneagram'* or a nine-sided star polygon is the Circle of TIME and NUMBER representing the Vertical plane and contraction. The Sun and 9 planets form a circle in which the 0 unites itself with the 9 like the mythic *Uroboros* - symbol of eternal time. The numbers are the octaves of densification that reflect the way Time condenses consciousness in an orderly spectrum from subtle to gross. As we can see, the 9, 6, and 3 form the basic triadic structure of this wheel. Ms. Norelli-Bachelet explains:

> *'In the circle of 9 we are describing, the most essential part is the 0 which unites itself with the 9, becoming the serpent biting its tail. Herein we see that Spirit and Matter are but different densifications or rates of vibration of the one all-pervading Energy, which to us is Absolute, God, Brahman.'*[58]
> **Patrizia Norelli-Bachelet, *The Gnostic Circle*, p.110**

> *"There are two poles of universal being. One is Spirit, the other is Matter. The essence of both is Consciousness in different states of vibration. In the Spiritual pole there is complete fusion of Time and Consciousness which we experience as Timelessness, or undifferentiated Time. In this state Time-energy vibrates at such an intense rate that it appears static and thereby lacking any element of periodicity or denseness. Hence it cannot produce any form or any division of Consciousness-substance into distinct crystallised objects in Space. As time slows down the densification of consciousness becomes greater, we experience the extreme lengthening of Time and a consequent solidification of energy where Time is experienced in its most clearly defined periods. Indeed, Matter is but Consciousness jelled into countless forms by the crystallising action of Time. Between these polar extremes lie the dream states and planes of subtle matter. As Time is thinned out, so too Matter is rendered more subtle. In these realms, our 'bodies' also thin out and we immediately*

learn that because material density is lacking, it is possible to pass through walls, doors, even other 'bodies'. Likewise, what appear to be vast distances are crossed in the batting of an eyelid, because rigid time barriers are not met in the more subtle planes of universal being. In the less dense planes all is more fluid, but, at the same time, a certain element is lost which is the most precious gift of our material universe. Here we actually progress, we evolve, we develop, we grow - because we are conscious of Time in its densest aspect. The price we must pay for this gift is the 'heaviness' of dense matter, in that the other worlds are contained in our own. Thus we carry the burden in our dense physical of all the other worlds." [59] **Patrizia Norelli-Bachelet, *The New Way*, Vol. 3 [underlining emphasis mine]**

For the Rishi, Time in the form of the Year was not something ephemeral. It was a container or womb upholding and gestating varying densities of form and planes of consciousness. When he sings to us of the months and the years being the secret of secrets...doors to certain eternal worlds, he is actually describing a synthesis of Time and Space wherein the most coveted secrets may be discovered. The Months are the 12 signs/divisions of the zodiacal circle and the Year contains the numerical scale of 9 connecting the planes of Spirit and Matter. These two circles in One unite the planes of Time and Space in a way that allows the realizer of the New Way to measure his orderly movement through the months and the years while observing the profound significance of the zodiacal archetypes and numerical octaves in his passage. The combined circle functions like an ancient Astrolabe allowing the individual to integrate these harmonies and navigate through his life with an awareness and orientation unequalled by any other means. Ms. Norelli-Bachelet's Gnostic Circle is that synthesis - a combination of the zodiac, (the occult circle which contains the knowledge of the evolutionary archetypes) - and the structural pattern of the solar system. In it we can see the *'thrice-seven, or twenty-one ecstasies'* which are the combined scales of 9 and 12.

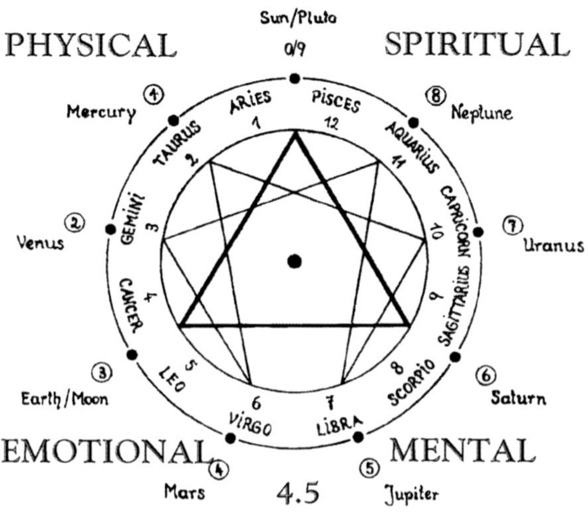

The Gnostic Circle

Patrizia Norelli-Bachelet
The Gnostic Circle
Samuel Weiser Inc, NY 1975

Ms. Norelli-Bachelet's discovery of the connection between the Rig Vedic hymns and the Zodiac by means of their common focus on the Year is ground-breaking. No one since the Vedic age has been aware of this essential correspondence, thus it can be said that no one has truly understood the Veda.

> *'...The Rig Vedic Myth is the same story of the evolution of human consciousness as that contained in the Zodiac.'*[60]
> **Patrizia Norelli-Bachelet, private correspondence regarding *Secrets of the Earth, 2009.***

It is like finding a *'Rosetta Stone',* where understanding the esoteric language of the one can shed light on the meaning of the other. When we read the account of the Biblical seers, Ezekiel and St. John the Divine, we discover the same zodiacal language as is contained in the passage from the Rig Veda that describes Vishnu's Three Steps. Anyone familiar with the zodiacal signs will immediately recognize the Bull, Lion, Eagle and Man as the Fixed signs of the zodiac which divide the wheel into four equal sections. There are levels upon levels of meaning in these ancient scriptures many of which have been fully revealed in Ms. Norelli-Bachelet's book *The Hidden Manna, the Revelation Called Apocalypse of Saint John the Divine.* [61]

As we move deeper into the knowledge of Twashtri's bowl and its four-fold planes of being we discover that his bowl, made into four, was to be the foundation or support of man's integral development. It was a circle/year divided into the four quarters of Being. As the individual lives through the months and years he begins to awaken to the existence of a hidden order (sign and number) behind his movement through time. A preexistent unity begins to reveal itself changing his outlook on reality and the world. Ms. Norelli-Bachelet has used the language 'Stable Constant' and 'Vedic Ecliptic Base' to describe this foundation upon which our personal churning takes place. It is analogous to Kurma the tortoise in the churning myth. And this is indeed the case. The Gnostic Circle provides us with stability and orientation as we live and relive these cycles of time with less and less resistance. It is a tool that allows for a progressive refinement of our consciousness and a true integration of the four quarters of our being. The Vedic sages taught that if man could learn to find himself within the harmonies of cosmic Time, he would regain his natural alignment and become the

Sun/center of his own Cosmos. We can define the New Way as the means by which this alignment is increasingly seen.

Becoming a Sun

As we apply this cosmological model to our natural lived experience, the real and true transformation begins to occur. A shift in consciousness is made, whereby the individual growth process undergoes a remarkable change: it begins to flower from a central point in all directions simultaneously and this gradually produces an integral perception or 'gnosis'. In the old spiritual paths the yogic process was linear, which meant that one set out from one point/chakra to reach a higher one. In a spherical growth process, the movement is outward from a central core, expansion from an axis encompassing the entire being. In this passage from the *Synthesis of Yoga*, Sri Aurobindo describes the features of our occult physiology in a way that helps us understand the necessity of this 'axial centering' and its true value.

> *'There are in fact two systems simultaneously active in the organisation of the being and its parts: one is concentric, a series of rings or sheaths with the psychic at the centre; another is vertical, an ascension and descent, like a flight of steps, a series of superimposed planes with the supermind-overmind as the crucial nodus of the transition beyond the human into the Divine.*
> *First, there must be a conversion inwards, a going within to find the inmost psychic being, and bring it out to the front, disclosing at the same time the inner mind, inner vital, inner physical parts of the nature. Next, there must be an ascension, a series of conversions upwards and a turning down to convert the lower parts.'*[62] **Sri Aurobindo, *The Synthesis of Yoga*, [underlining emphasis mine]**

Sri Aurobindo 1872 - 1950

Clearly the Axis and Plane are not the exclusive property of Galaxies, Solar systems and Planets; they are also the preeminent feature of human existence. Sri Aurobindo describes the Axis within the individual as a vertical system like a flight of steps leading to the Supermind. The Plane is shown as a series of concentric rings or horizontal sheaths with the psychic-being or Soul at the Center. Once we have secured that ecliptic base from which we can begin the centripetal journey inward, it is simply a matter of progressive refinement and seeking a more perfect axial alignment until we reach the inner chamber of the Soul.

The movement inward toward the center is not without gifts. It is met by an ever increasing revelation compared by the Rishis to a slowly rising Sun. The embodiment of this emerging light in the Vedic pantheon is known as *Usha*, the Dawn goddess:

> *'The journey is led by Usha the Dawn, the noble and active Goddess who hath emerged from the darkness... Dawn on her shining chariot is resplendent...eager for conquest, with bright sheen she cometh making apparent the lovely treasures*

> *which the darkness had covered ...Fair as a bride embellished by her mother thou showest forth thy form that all may see it.'*[63] **Rig Veda: Mandala 1, Hymn 123**

In '*The Life Divine'*, Sri Aurobindo gives us a vivid image of what Usha reveals to the fearless seeker of the Vedic Sun:

> *'...We perceive a graduality of ascent, a communication with a more and more deep and immense light and power from above, a scale of intensities which can be regarded as so many stairs in the ascension of Mind or in a descent into Mind from That which is beyond it. We are aware of a sea-like downpour of masses of a spontaneous knowledge which assumes the nature of Thought but has a different character from the process of thought to which we are accustomed; for there is nothing here of seeking, no trace of mental construction, no labour of speculation or difficult discovery; it is an automatic and spontaneous knowledge from a Higher Mind that seems to be in possession of Truth and not in search of hidden and withheld realities. One observes that this Thought is much more capable than the mind of including at once a mass of knowledge in a single view; it has a cosmic character, not the stamp of an individual thinking. Beyond this Truth-Thought we can distinguish a greater illumination instinct with an increased power and intensity and driving force, a luminosity of the nature of Truth-Sight with thought formulation as a minor and dependent activity. If we accept the Vedic image of the Sun of Truth, - an image which in this experience becomes a reality, - we may compare the action of the Higher Mind to a composed and steady sunshine, the energy of the Illumined Mind beyond it to an outpouring of massive lightnings of flaming sun-stuff...'*[64] **Sri Aurobindo, *The Life Divine*, pp.277-78 [underlining emphasis mine]**

In this extraordinary passage Sri Aurobindo reveals some central features of this New Way of knowing. First, it has a Cosmic character and secondly, '*There is nothing here of seeking, no trace of mental construction, no labour of speculation or difficult discovery'.* It is a movement from truth to higher truth in the ascent to the Supermind. At the same time this luminous consciousness flows downward to convert and purify the lower parts of the being allowing one's higher possibilities to be developed.

When we contrast this New Way of knowledge to the fruitless speculations of the modern scientific community, it confirms the underlying error of their approach. They will never arrive at a unified truth by following the circling tracts of mind. Until scientists begin to recognize the concept of the axis/center, its nature and specifics, they are destined for repeated failure. And this is where we encounter the likelihood of some very grave problems. While Sri Aurobindo wrote of the Cosmic character of the supramental gnosis, the Mother revealed an even deeper quality of that cosmic alignment that the New Way confers on its seeker.

> *'A person who has undergone the process of transformation effects in turn, a transformation of events around them. It is not magic or an occult power, it is rather that his or her being present, is in itself sufficient to create a 'cosmos' or harmonious field which could be nowhere manifest without his or her presence.'*[65] **The Mother**

As the Mother reveals, the nature of the New Seeing is not a passive observation. It is a formative gnosis that sees the truth because one has achieved an axial alignment that orders his field or plane. Having witnessed this phenomenon first hand in 2005 at the Aeon Centre of Cosmology, [B] I can assure the reader that this is no idle claim. And if we accept that this is indeed true, we may conclude that the converse is also true. So long as particle physicists continue to project their own misaligned and center-less condition onto what they see and experiment upon, they will invariably come up with the wrong conclusion. The problem lies in the fact that physicists now believe that the key to a Grand Unified Theory can only be found by going deeper into the immense densities, pressures and temperatures which lie at the quantum singularity. To achieve this goal they have spent over $10 billion to develop a technology known as the Large Hadron Collider (LHC), which is the world's largest and highest-energy particle accelerator. This machine has the power to smash elementary particles into each other at unthinkable speeds to try and recreate the conditions of the 'big bang'.

We know of course that the mysteries of creation, the order, laws and forces lie beyond the event horizon of matter in the Zero-Womb. Consequently, the truth they seek cannot be found in sub-atomic particles no matter how long or hard they search. But their naïve willingness to tamper at the threshold of the awesome forces that created the universe exposes us all to incredible danger. The question to consider is this: What kind of careless risks are arrogant scientists willing to take to try and justify this patently untenable position? Since it is moving in a direction counter to what a supramental science would consider prudent, the odds of an unexpected accident grow stronger every day.

The Large Hadron Collider at CERN - the European Organization for Nuclear Research - bears a certain similarity to the legendary Crystal of Atlantis known as the *Tuaoi* or 'Firestone'. It was a huge six-sided cylindrical crystal which was said to be able to capture solar, lunar, atmospheric and telluric energies and focus them in such a way that produced tremendous power. According to the renowned psychic Edgar Cayce, the crystal amplified the powers of the Sun itself to create a Ray that could generate energy from the disintegration of the atom. As the arrogant priests of the Great Crystal became greedy for more power, said Cayce, the energies of the Firestone were tuned to higher and more destructive frequencies until in their carelessness they finally destroyed their entire civilization. The story of Atlantis has special meaning for our modern world in the sense that we too are vulnerable to an arrogant science that has become disconnected from the sacred Earth. The only thing that can reverse this dangerous threat is a willing embrace of a sweeping change in consciousness.

The Descent of the Supermind

> *'Anything that disturbs the Inertial is, for Inertia, a catastrophe. In the world, the earthly world (it is the only one I can speak of with competence ...), in the earthly world, for Inertia (which is the basis of the creation and is necessary to fix, to concretize things), anything that disturbs it is a catastrophe. That is to say, the advent of Life was a monstrous catastrophe, and the advent of intelligence in Life another monstrous catastrophe, and now the advent of Supermind is the final catastrophe! That's how it is. And for the unenlightened mind it really is a catastrophe! .... The whole system spends its time rejecting and rejecting all that comes. ...* '[66] **The Mother, 28 November 1964, *The Mother's Agenda*, Volume 5**

As she approached her 92nd birthday, the Mother began to SEE a form on the plane of truth-consciousness that would be her greatest gift to mankind. It was a plan for a Vedic Temple, a masterpiece of sacred architecture whose major characteristic was Unity. The Mother's plan was a vision of mathematical harmony joining the Vertical Axis and Horizontal Plane at a luminous and perfect Center. It established a new harmony between Time and Space and expressed this truth in objective metric measurements. It was a representation of the Soul of the Earth.

When a seer as impeccable as the Mother sees a supramental form and accurately translates its laws and principles, it allows them to manifest on the Earth. And when they come down, they carry a formidable power of transformation because they are the symbols indicating the descent of a new consciousness into the Earth plane. *'And when this higher power is coming down,'* said the Mother, '*it comes as an absolutely luminous and perfect organization which one can see if he has the vision...it presses on matter and everything begins to seethe and resist.'* When Sri Aurobindo was asked about the signs of the imminence of the Supramental Descent, he replied:

> *'...I find that the more the Light and Power are coming down the greater is the resistance. It is always the sign that whenever the higher Truth is coming down, it throws up the hostile vital world on the surface, and you see all sorts of abnormal vital manifestations, such as an increase in the number of persons who go mad, weather disturbances, and earthquakes....'*[67] **A. B. Purani, *Evening Talks with Sri Aurobindo*, The Second Series, 15-8-1925**

The evolutionary pressures that accompany this new luminous order are relentless because they are pressing for the release of the energy imprisoned in our old mental and vital forms, thus allowing it to re-form at a higher poise of organization. The world is being churned so that this can occur. Those who can align themselves to receive this force through the unveiling of an individual Center or Axis will escape these destructive pressures and wake up to a brave new world. But the typical human being, the average worker, housewife or professional that we meet every day of our lives has no such Center. They operate with a completely erroneous sense of self. While they may appear highly intelligent and successful within the framework of our collective social fabric, in actuality their lives orbit a meaningless void, a hub of nothingness. Their day to day experiences are simply a repetition of purposeless atavistic drives and indulgences inspired by the separative ego. There is usually a mass of internal contradictions brought about by an unresolved dissonance between the mental, emotional and physical parts of the being, not to mention the spiritual. Such individuals frequently act against their own self interest and the wellbeing of others without the slightest understanding of what they are doing. For all practical purposes he or she is asleep.

Indeed, Plato's cave is not exclusively populated by misfits and miscreants. It holds equal numbers of scientists, doctors, lawyers, teachers, housewives, and even captains of industry. And if our typical individual happens to live in the Western, upwardly mobile consumer culture he or she is assaulted on all sides with the unending demands of materialism, a default lifestyle which most of their fellow humans have warmly embraced. If there is any concession given to the non-material or 'spiritual' side of their life, it consists, at best, in attending superstitious religious services where the faithful are promised salvation and fulfillment in an after-death heaven. Or, if they happen to be more scientifically inclined, they might pick up the latest publication by the world famous physicist, Stephen Hawking, in which he authoritatively states that *'[God] did not create the universe...given the existence of gravity, the universe can and will create itself from nothing...Spontaneous creation is the reason why there is something rather than nothing ... why the universe exists, why we exist.'* [68] Hawking believes that our universe is just one among billions of possible universes, and that it is statistically unremarkable that one would have the right combination of cosmological values to support conscious life as it exists on Earth. According to the person that many people believe to be the 'smartest scientist in the world', our universe is a meaningless, purposeless coincidence of physics that has no creator, organizer or sustainer.

When the two great pillars of civilization, Religion and Science reach a point where they both agree that the world has no meaning or purpose, the implications of that insidious nihilism can be devastating. An individual born under this influence has no sense of order and orientation. For him, there is no central upholder, no replenishing of energy, no organizing force and consequently no vision of purpose or awareness of a destiny. The mental and emotional energies of someone in this disposition are fragmented, scattered and poised to collapse. Without that Center/Soul/Axis at the core of one's being a person simply cannot withstand the contracting pressures that such a meaningless life brings to bear. The inevitable psychological response of an individual under these conditions is **escape or collapse**. Incidents of both of these tactics along with growing numbers of people going mad are increasing exponentially all over the world.

Statistics indicate that depression is rapidly increasing in industrialized countries, and no one knows why. According to the U.S. Centers for Disease Control 25% or almost 80 million Americans suffer from mental illness. The CDC estimates that one in 10 American adults, or 31 million people are suffering from depression; in Great Britain estimates are as high as one in six. The National Health and Nutrition Examination Survey estimates that 31 million people are on anti-depressant drugs and over 30 million Americans are addicted to illegal drugs and alcohol. Another 80 million are borderline users abusing substances in a way that threatens their health and safety. But one of the most shocking statistics is from the World Health Organization. The WHO has just reported that more than 350 million people suffer from depression globally and it has predicted that in 8 years, 2020, mental illness will be the second leading cause of death in the world.

What we can deduce from these shocking statistics is that something is producing measurable changes in the mental and emotional wellbeing of large numbers of people. In order to mask the painful effects of this pressing influence, there has been an explosive increase in the use of pharmaceutical drugs. In the 1970s and 1980s Pharmacological Science began to manufacture tranquilizers and anti-depressants on a massive scale to help people cope. Since they are no more than a band-aid on a much deeper issue this has resulted in some disastrous side effects. In a growing number of users the same anti-depressants that originally provided relief are fueling an epidemic of drug-induced mental illness making their victims even less capable of coping than they were before. Another irrefutable indicator of collapse is the explosive increase in cases of Dementia or Alzheimer's disease, which are inescapable symbols of the dissolution of the mind.

In the last decade we have seen a quickening of these evolutionary pressures in the form of social and economic collapse. In the U.S. we have seen an epidemic of unemployment, poverty and homelessness, and the looming collapse of the fiat currency and financial system. In Europe conditions are even worse including the financial insolvency of many sovereign nations. What must be understood is that these events are not a periodic economic aberration like a 'Great Depression'. They are the result of a relentless and purposeful descent of force that will continue unabated until the job is done. The Mother explains:

> *'No sign will enlighten those whose eyes remain closed. But for those whose look is clear, darkness itself becomes a sign.*

For do they not know that the night grows darker as the dawn approaches? Nevertheless, we will give to everyone a means of discernment: When all moves and quakes, when a shiver passes through the people, awakening those who had been plunged in sleep for centuries and threatening thrones, when that which seemed immovable begins to waiver, when the proudest and most solid constructions shake on their bases and threaten ruin because the very foundation of things is displaced, then can be seen the advent of one whose superhuman steps make the earth tremble. '[69] **The Mother**

You will not find a more precise description of the conditions afflicting our modern global society. What is being displaced in both the Individual and the Mass is the foundation of our binary mental perception of reality. When that which we have believed to be true for most of our lives begins to waiver and threatens ruin we suddenly find ourselves faced with a painful and frightening disorientation. For most this can be a rude awakening especially if they have no understanding of the Unitary nature of reality and the means to adapt to its higher laws. It's like waking up in a foreign country where you do not know the customs and cannot speak the language. Our usual support systems are unable to help because most have no idea what is going on and cannot accommodate the demands of these evolutionary pressures. In other words, no one is going to save us. They don't know how. We must discover a way to save ourselves.

'This struggle, this conflict between the constructive forces of an ascending evolution, of an increasingly perfect and divine realization, and the more and more destructive forces – powerfully destructive forces of an uncontrollable madness – is becoming more obvious, unmistakably visible, and it is a kind of race or battle as to which will be first to reach its goal. '[70] **TheMother [underlining emphasis mine]**

Ms. Norelli-Bachelet's New Way is truly the only way to engage this *'Ascending Solution'* because she provides both the knowledge and the means for us to align with a new foundation of reality and become willing instruments of its perfection. To properly understand this possibility we must thoroughly reconsider what we have come to believe about the nature of Spirituality.

An Integral Spirituality

Over the millennia our experience of Spirituality has gone through many changes and forms. In the ancient past Goddess worship was the norm throughout much of the world. The most recent example of this revered tradition was the cult of Demeter and Persephone based at Eleusis in ancient Greece. The myth is permeated by the transformative unity of Mother and Daughter, Demeter and Kore, the Earth and her Soul. The unity of Spirit, Nature and the Individual was the central content of the Eleusinian mysteries which came to an end in 396 BCE when the state took over control of the mysteries. In the millennia that followed spirituality began to move away from the organic wholeness of spirit and matter that the Goddess embodied and toward a more one-sided spiritual view that regarded the realization of an utterly Transcendent Absolute as the highest possible attainment. The sacred experience of a magical enchanted world soon gave way to notions that life on Earth was simply a veil of suffering, the 'fall of the soul' from which the seeker must escape in order to attain salvation. Concepts of illusionism that considered the world of matter as unreal came into vogue along with religious notions of heavens, nirvanas, or paradisiacal afterlives. Out of this denial of the Earth emerged a rule-bound, Patriarchal, transcendent model of reality that consigned all the domains of the sacred Feminine (the Cosmos, Time, Matter, Destiny, Meaning, and Purpose) to the waste heap of history. The primary feature of this new mental/religious model is its warm embrace of dualism; the irreconcilability of Spirit and Matter. We can see this binary view played out in our modern culture in the worship of a transcendent God who is distinct from and not a part of His material creation. Over the centuries our unquestioned acceptance of this demonstrably false view has obliterated our awareness of the awesome sacredness of Nature and removed all restraints on its unconscionable exploitation by a reckless and soul-less science.

For millennia seekers have been confronted with this Binary perception of reality: Spirit and Matter, Masculine and Feminine, Shiva and Shakti, Yin and Yang. But in order to SEE what Really IS we must stop looking at the world through the narrow aperture of an old bi-polar creation. This means widening our perspective on both nature and ourselves. Ms. Norelli-Bachelet has shown us that the Supramental creation is not Binary, it is an integral spectrum of consciousness from spirit to matter, from Divine to human, including many intervening layers of reality and being that find their ultimate reconciliation in the human Soul. With the advent of Supermind the Binary has given way to the Tertiary, thesis and antithesis to Synthesis. This differentiation of the Divine consciousness into Three aspects rather than the usual Two is explained by Sri Aurobindo:

> *'While the Divine is One, it is also manifold...It is at once Transcendental, Cosmic and Individual. By knowing the eternal unity of these three powers of the eternal manifestation, God, the Cosmos and the Individual self, and their intimate necessity to each other, we come to understand*

<u>existence itself</u>... '.[71] **SriAurobindo, *The Synthesis of Yoga, p. 439* [underlining emphasis mine]**

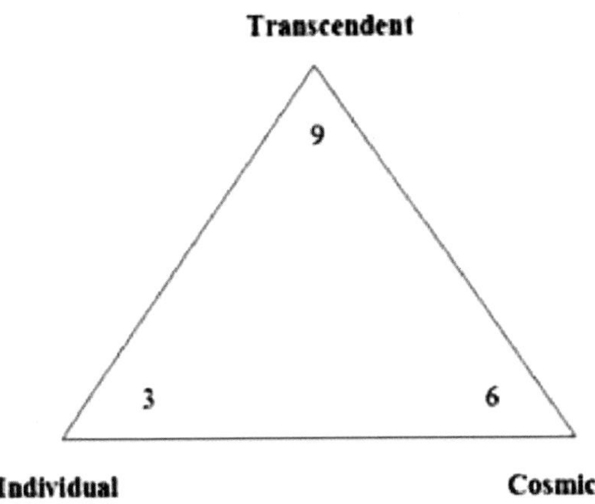

'The distinction between the Transcendental, the Cosmic, the Individual Divine is not my invention, nor is it native to India or to Asia -- it is, on the contrary, a recognised European teaching current in the esoteric tradition of the Catholic Church where it is the authorised explanation of the Trinity, -- Father, Son and Holy Ghost ...[In] the practice of yoga there is a great dynamic difference in one's way of dealing with these three possible realisations. If I realise only the Divine as that, not my personal self, which yet moves secretly all my personal being and which I can bring forward out of the veil, or if I build up the image of that Godhead in my members, it is a realisation but a limited one. If it is the Cosmic Godhead that I realise, losing in it all personal self, that is a very wide realisation, but I become a mere channel of the universal Power and there is no personal or divinely individual consummation for me. If I shoot up to the transcendental realisation only, I lose both myself and the world in the transcendental Absolute. If, on the other hand, my aim is none of these things by itself, but to realise and also to manifest the Divine in the world, bringing down for

> *the purpose a yet unmanifested Power, -- such as the supermind, -- A HARMONISATION OF ALL THREE BECOMES IMPERATIVE. I have to bring it down, and from where shall I bring it down -- since it is not yet manifested in the cosmic formula -- if not from the unmanifest Transcendence, which I must reach and realise? I have to bring it into the cosmic formula and, if so, I must realise the cosmic Divine and become conscious of the cosmic self and the cosmic forces. BUT I HAVE TO EMBODY IT HERE, -- OTHERWISE IT IS LEFT AS AN INFLUENCE ONLY AND NOT A THING FIXED IN THE PHYSICAL WORLD, AND IT IS THROUGH THE DIVINE IN THE INDIVIDUAL ALONE THAT THIS CAN BE DONE.* '[72] **Sri Aurobindo, *Letters on Yoga*, SABCL Vol. 22/24, p. 509-10 [capitalization and underlined emphasis mine]**

> '*...This triple movement is the whole key of the world enigma ...The first founds the inalienable unity of things, the second modifies that unity so as to support the manifestation of the Many in One and the One in Many; the third further modifies it so as to support the evolution of a diversified individuality, which, by the action of Ignorance, becomes in us at a lower level, the illusion of the separate ego.* '[73] **Sri Aurobindo, *The Life Divine*, Chapter XVI, 'The Triple Status of Supermind'**

This Triple aspect of the Divine was discovered over 5,000 years ago by the Vedic and Egyptian civilizations, and later surfaced in our Christian notion of the Trinity of God. It is based on the *'Law of Three'* - a tri-fold point, which lies at the core of matter and holds everything in a superb harmony of the One and the Many. In the Supramental Cosmology this principle is expressed numerically as 9 - 6 - 3. We can see these qualitative divisions being embodied by the three individuals discussed in this study: Sri Aurobindo – the Transcendent 9, the Mother – the Cosmic 6, and Ms. Norelli-Bachelet – the Individual 3.

The 6 and the 3

Through the Supramental philosophy of Sri Aurobindo '9' we have been given a Transcendent Overview, the scope of which has not been attained by any other known thinker. But what good is this sublime knowledge if it cannot be brought down and realized in our everyday lived experience? Failing that it is just another mental Belief. According to Sri Aurobindo this is precisely the point and the Herculean task of bringing the knowledge down can only be accomplished through the feminine principles, the Cosmic '6' and Individual '3' because ultimately, *'IT IS THROUGH THE DIVINE IN THE INDIVIDUAL ALONE THAT THIS CAN BE DONE'.* The ground breaking importance of Sri Aurobindo's statement can be seen in juxtaposition to an old spirituality that emphasized the Transcendent 9 and denied the contribution of the 6 and 3. This is precisely why our traditional religious notions of the Spirit remain 'up there' and based on what we believe God wants, instead of the Divine Will being brought down and **made visible** in the physical world of matter. This is not possible with the 9 alone and the unfortunate result of an exclusively transcendental spirituality has been a civilization torn apart by competing mental/religious *beliefs* seeking to dominate the minds and hearts of the faithful. Appeasement and compromise between these old patriarchal religions is not the answer to our pressing global problems. The only solution is a non-speculative, integrating *knowledge* of the Spiritual that rises above all mental/religious distinctions.

The precious gift of the Supramental formula 9-6-3-0/1 allows us to articulate just such a knowledge by following the descent of the *Transcendent Divine* from its lofty heights down and through the cosmic and planetary harmonies before reaching a highly specific manifestation in the Individual as the *Immanent Divine*. All spiritual literature has acknowledged this process in one form or other resulting in the rare and iconic 'Sons and Daughters of God'. With the descent of the Supramental Consciousness the *Immanent* realization is not only possible for all, it is an evolutionary necessity for the advent of a Life Divine on the Earth. We are indeed fortunate to have such impeccable examples of this descent so that we may follow its progress in the lived experience of the 6 and 3 of the line, the Mother and Ms. Norelli-Bachelet. One of the most stunning examples of the Cosmic realization can be found in the Mother's 1915 letter to Sri Aurobindo. She sent the following description of an experience she had that seems to have been her investiture as the Cosmic Principle, the 6 in the work that lay ahead for her with Sri Aurobindo as the Transcendent 9:

> *'The entire consciousness immersed in divine contemplation, the whole being enjoyed a supreme and vast felicity.*
>
> *'Then was the physical body seized, first in its lower members and next the whole of it, by a sacred trembling which made little by little even in the most material sensation all personal limits fall away. The being progressively, methodically, grew in greatness, breaking down every barrier, shattering every obstacle that it might contain and manifest a power which increased ceaselessly in immensity and intensity. It was as if a progressive dilation of the cells until there was a complete identification with the earth: the body of the awakened consciousness was the terrestrial globe moving harmoniously in ethereal space. And the consciousness knew that its global body was thus moving in the arms of the universal Personality, and gave itself, it abandoned itself to Her in an ecstasy of peaceful bliss. Then it felt that its body was absorbed in the body of the universe and one with it; the*

consciousness became the consciousness of the universe, in its totality immobile, in its internal complexity moving infinitely. The consciousness of the universe sprang towards the Divine in an ardent aspiration, a perfect surrender, and it saw in the splendour of the immaculate Light the radiant Being standing on a many-headed serpent whose body coiled infinitely around the universe. The Being in an eternal gesture of triumph mastered and created at one and the same time the serpent and the universe that issued from it, erect on the serpent he dominated it with all his victorious might, and the same gesture that crushed the hydra, enveloping the universe, gave it eternal birth. Then the consciousness became this Being and perceived that its form was changing more and more; it was absorbed into something which was no longer a form and yet contained all forms, something which, immutable, sees – the Eye, the Witness. And what It sees, is. Then this last vestige of form disappeared and the consciousness was absorbed into the Unutterable, the Ineffable.

'The return towards the consciousness of the individual body took place very slowly in a constant and invariable splendour of Light and Power and Felicity and Adoration, by successive gradations, but directly, without passing again through the universal and terrestrial forms. And it was as if the modest corporeal form had become the direct and immediate vesture, without any intermediary, of the supreme and eternal Witness.'[74] **The Mother, 1915 letter to Sri Aurobindo**

The Mother 1878 - 1973

In a letter dated Dec. 31, 1915, Sri Aurobindo confirms that the Mother's experience was entirely Vedic, and of an order that would not be recognized by today's yogis.

> *'The experience you have described is Vedic in the real sense, though not one which would easily be recognised by the modern systems of Yoga which call themselves Vedic. It is the union of the "Earth" of the Veda and Purana with the divine Principle, an earth which is said to be above our earth, that is to say, the physical being and consciousness of which the world and the body are only images. But modern Yogas hardly recognise the possibility of a* ***material*** *union with the Divine.'*[75] **Sri Aurobindo, *Letter to the Mother*, The Mother's Collected Works, *1915.* [bold and underlined emphasis mine]**

Sri Aurobindo's reply makes the point that the modern systems of yoga that call themselves Vedic would not have recognized her experience. This is because they too had been stuck at the Transcendent 9 and could not even imagine the possibility of an Integral realization.

Many of the 'spiritual' experiences that the Mother had throughout Her 95 years are recorded in her *Agendas*. It must be acknowledged that She was one of the greatest adepts of the Age. The accounts of Her explorations of the inner worlds and their multiple dimensions have no equal in the world's spiritual literature. Gifted with extraordinary abilities, the Mother took every opportunity to delve into the connections between the Spiritual and Material worlds. Of particular interest to Her was the possibility of some kind of reconciliation between spirituality and science.

> *'These positions, the spiritual and the "materialist" if one may call it so, that are believed to be exclusive (exclusive and unique, so that one denies the value of the other, from the viewpoint of Truth), are insufficient, not only because they do not admit each other, but because even admitting the two and uniting the two does not suffice to solve the problem. There is something else –* ***a third thing*** *which is not the result of these two, but something that is yet to be discovered, which will probably open the door to the total Knowledge... .'*

> *'For a very long time it seemed to me that if one made a perfect union between the scientific approach carried to its extreme and the spiritual approach carried to its extreme – its realisation – , if one joined these two, one would find, one would obtain naturally the Truth one seeks, the total Truth. But with the two experiences that I had, the experience of the external life (with the universalisation, impersonalisation, with all the yogic experiences that one can have in the material body) and then the experience of the total and perfect union with the Origin, now that I have had these two experiences and there has occurred something – which I cannot describe now – I know that the knowledge of the two and the union of the two are not sufficient; there is a third*

thing in which these two terminate and it is ***this third thing which is in the making****, in the process of working itself out. It is this third thing that can lead to the Realisation, the Truth we seek...*

'I arrived, by yoga, at a certain kind of relation with the material world based on the notion of the fourth dimension (inner dimensions that become innumerable in yoga) and I made use of this attitude and this state of consciousness. I studied the relation between the material world and the spiritual world with the sense of inner dimensions and by perfecting the consciousness of these inner dimensions – that had been my experience before the last one.

'Naturally, for a long time, there was no longer any question of three dimensions – that belonged absolutely to the world of illusion and falsehood. But now it is the use of the sense of the fourth dimension with all that it entails which appears to me as superficial! I do not find it anymore, the thing is so strong. The other, the three dimensional world is absolutely unreal; and the other appears, how to say, conventional. It is as it were a conventional translation to give you a certain kind of approach.

'And as for saying what it is, the other one, the true position? ...It is so much beyond all intellectual states that I am unable to formulate it.

'It is that something that we are searching for. Perhaps not merely searching for, but building.

'But the formula will come, I know. But it will come in a series of lived experiences that I have not yet had.'[76]**The Mother, *The Mother's Agendas*, 1962**

The Mother's statements give us an extraordinary glimpse into the difference between an old static realization of the spirit 'up there' and the dynamic process of its Becoming here in the material world. This is the signature feature of the 6 and 3 in the line, to embody the truth in the very fabric of their lived experience. Ms. Norelli-Bachelet confirms:

> *'Indeed, those experiences did begin eight years later in 1970 when the Mother 'saw' the chamber[of Her Temple], measured it and gave the plan to her entourage to execute as the centre of Auroville, a plan which was never implemented but rather entirely altered to become unrelated at all to her original. But this was just the beginning. In 1971, the ninth year after the above talk, the very precise formula she anticipated began to take shape through the third in Sri Aurobindo's line whereby those 'conventional' three dimensions underwent a radical transformation. From 1976 onward, through the lived experiences the Mother mentioned, the new position, that 'third thing' manifested in full. When that happened indeed the perception of dimensions shifted to a new formula that better explains the reality of our world – both here on this side of the material demarcation, and the other in what is considered beyond or before the singularity of physics. The perception now, as of 1983 as formulated in the new cosmology (see* The New Way, *Volume 3, Aeon Books, 2005) is three dimensions of Time, the fourth of which is Space – that is, the fourth is the Point (of space) ...'* [77]
>
> **Patrizia Norelli-Bachelet, 'The Vedic Altar', 2012, p. 22**

The 'Third Thing'

One of the most important principles in philosophy is the concept of dialectics as set forth in the early 1800s by Georg Wilhelm Friedrich Hegel. It is an interpretive method in which the contradiction between a proposition *(thesis)* and its opposite *(antithesis)* is resolved at a higher level of truth *(synthesis)*. Hegel's method is a perfect example of 'the Law of 3' in which a triadic logic is used to resolve the opposition between polar extremes in order to achieve a higher synthetic unity. The ancient Alchemist expressed this same dialectic process in the numerical formula, *One becomes Two, Two becomes Three, and out of the Third comes the One as the Fourth.*[78] This is the occult process through which the Transcendent 9 becomes the Immanent 1, or as Ms. Norelli-Bachelet has expressed it, 9-6-3-0/1, the formula of the Supermind - '*Three dimensions of Time, the fourth of which is Space*'. Between the 9 and the 1 lie the two feminine powers, the Cosmic and the Individual, the 6 and 3. And it is they who create forms for the formless, living symbols that resolve the age old contradictions and construct a bridge between the spiritual and material planes. As the Cosmic principle, the Mother encapsulated that bridge in the form of a Vedic Temple which was the greatest feat of sacred architecture yet to be realized upon the Earth but it was Ms. Norelli-Bachelet, the 3, who gave it Movement.

Like Hegel, Ms. Norelli-Bachelet understands that the first casualty of polar oppositions is movement, dynamism, progress. We can see this inertia and lack of direction so clearly in the rigid political partisanship presently afflicting the U.S. Congress. The work of the 3, therefore, is not only to discover a New Way to reconcile the bi-polar relationship between the Spiritual and Physical worlds but to restore the natural action and dynamism between them that will allow us to move beyond the inertia and stagnation of the past. Since Time and Movement are one and the same, it is clearly the key to a new and dynamic creation.

When we think of 'Synthesis' in the conventional sense it means the resolution of opposites, a combination of two or more entities that together form something new. But when Time is used as the basis of synthesis as Ms. Norelli-Bachelet has done, it provides the context for an all-encompassing knowledge that not only resolves duality but puts order to the world and sets everything in its rightful place. It gives us the means to SEE a preexisting Unity behind what appear to be random events, and allows us to understand the purposeful progression of history. Sri Aurobindo explains:

> *'From the beginning the whole development [of the universe] is predetermined in its self-knowledge and at every moment in its self-working - it is and moves to what it must be by its own original Truth, and will be at the end that which was contained and intended in its seed.'*[79] **Sri Aurobindo, *The Life Divine*, 'The Supreme Truth-Consciousness'**

Imagine having the benefit of a revolutionary new perspective that would allow you to look back over the years of your own personal history and suddenly have it fall into a seamless and meaningful order. A single glimpse would not only destroy the illusion of the ego's control but would provide the means to move into the future aligned with a completely new sense of purpose and understanding. With Ms. Norelli-Bachelet's Cosmology we now have the means to do just that. It gives us the ability to restore movement, meaning, purpose, dynamism, and knowledge to our individual and collective lives. These are the very qualities of the sacred feminine which for thousands of years science and religion have tried to stamp out of existence.

Indeed, the Supramental Time formula is the 'Third Thing'. And with the work of the Mother and Ms. Norelli-Bachelet, the 6 and the 3, we have for the first time the means of a perfected knowledge, and the basis for a new and constructive action in the world.

The Individual - 3

One of the problems of mainstream religions is that they are all based upon the subjective spiritual experiences of the original teacher, prophet or realizer. In virtually every case these experiences remain unverifiable and ineffable. It remains for the follower to take their experiences of God completely on faith, even though most of these accounts are thousands of years old and border on myth. What has evolved from our indiscriminate willingness to believe in things not seen has been a broad attitude of ecumenism in which all religions are now regarded as relative equals. No one faith is any better, or more true than another. This has led to the erroneous conclusion that, 'all truth is relative'. The pitfalls of religious relativism are self-evident when it comes to where we draw the line as to what claims are valid and what are not. But beyond this there is another, more important issue. Contrary to what these 'relative' religions believe, Spirituality is not a static and unchanging reality, it is a progressive movement, the same as the biological evolution, constantly revealing new and fuller aspects of the Divine Consciousness. With the advent of the Supermind we finally have the means to move beyond the chaos of relativism and into a collective realization of non-speculative Absolute truths, especially of Time and Space. As we follow the Supramental descent from the Transcendent to the Cosmic down to the Individual we will observe an ever-increasing ability to know and describe that Absolutism. Sri Aurobindo has given us an incredible Transcendent overview which the Mother has taken up and brought into a clearer focus by creating new forms for the formless. But it isn't until the 3rd level that these details become completely revealed because the nature of the 3 is *precision in truth.* The ability to SEE with such detail comes about because the action is observed from the most intimate dimension the human consciousness can experience, the Center or individual Soul.

The Individual 3 is the personification of the human Soul, that Point/Axis/Center from which all may be known. And if the Transcendent 9 is Timeless, the Soul 3 has the most intimate relationship with Time because it is the source of her Gnosis. To better understand this relationship between Time and the Soul we need only listen to the testimony of the Upanishadic sages:

> *'A being, the size of a thumb, like a flame without smoke, dwells deep within the heart.* ***He is the lord of time, past and future, the same today and tomorrow****. Having attained him, one fears no more. He, verily, is the immortal Self...'.*[80] ***Katha Upanishad***

In this brief but revealing passage the sage tells us that the Divine Self or Soul, located in the lotus of the Heart, is the *'lord of time, past and future, the same today and tomorrow'*. What exactly does he mean? Sri Aurobindo explains that one of the primary attributes of the Soul is the triple time knowledge, *'Trikaladristi'*, held of old to be a supreme sign of the seer and the sage, not as an abnormal power, but as a normal way of time knowledge. In this condition, the Seer enjoys an experience of 'Simultaneous Time' which resolves the apparent duality of the timeless Infinite and that of the Infinite deploying itself and organizing all things in time. The Mother explains:

'In a certain state of consciousness (I no longer remember what he calls it – I think it's in the 'Yoga of Self-Perfection'), one is perfectly identified with the Supreme, not in his static but in his dynamic aspect, the state of becoming. In this state, everything is already there from all eternity, even though here it gives us the impression of a becoming. ***And Sri Aurobindo says that if you are capable of maintaining this state, then you know everything: all that has been, all that is and all that will be – in an absolutely simultaneous way.*** *'*[81]
The Mother, Mother's Agenda, 18 April 1961

'...The Being can have three different states of its consciousness with regard to its own eternity.... ***The second is its whole-consciousness of the successive relations of all things belonging to a destined or an actually proceeding manifestation, in which what we call past, present and future stand together as if in a map or settled design or very much as an artist or painter or architect might hold all the detail of his work viewed as a whole, intended or reviewed in his mind or arranged in a plan for execution; this is the stable status or simultaneous integrality of Time.*** *This seeing of Time is not at all part of our normal awareness of events as they happen, though our view of the past, because it is already known and can be regarded in the whole, may put on something of this character; but we know that this consciousness exists because it is possible in an exceptional state to enter into it and see things from the view-point of this simultaneity of Time-vision. '* [82] **Sri Aurobindo, *The Life Divine*, p. 380**

Thea (PatriziaNorelli-Bachelet)

As the Individual Soul principle, Ms. Norelli-Bachelet has been invested with a unique realization of Time similar to what the Mother experienced in her investiture as the Cosmic principle. It has allowed her to SEE with such a totality of vision that she has been able to 'rectify' or organize with a meaningful precision the whole scope of human history. Her 'Map of the 12 Manifestations' follows the evolution of the Earth through the zodiacal ages by means of a phenomenon known as the 'Precession of the Equinoxes'. This provides us with a cosmological key to understand the order and organization of both ancient and modern civilizations. In this Map we will see yet another example of the Law of Correspondence and Equivalence which is based on the realization that both the development of the Individual and our Collective movement through history are simultaneous and interrelated expressions of the same evolutionary process. Both are under the impeccable control of *Mahakala* – the Time Spirit. What we will discover in the Map of Manifestations is a proportionate relationship between the four quarters of the human cycle and the

'Manifestations' of history. Both are based upon the division of four marked off by the Solstices and Equinoxes.

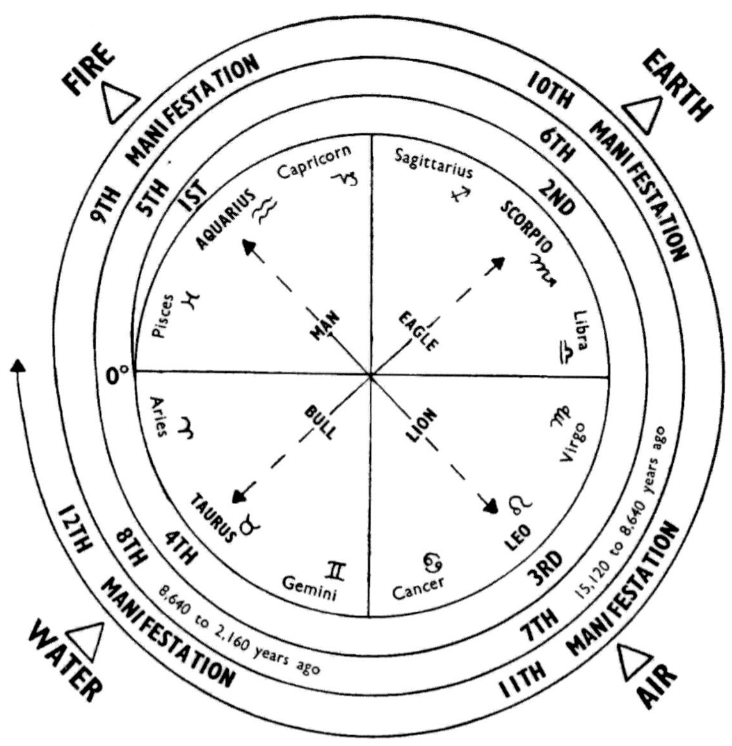

THE 12 MANIFESTATIONS
Map of the Evolutionary Ages

Out of her profound Seeing of the archetypal dynamics of historical Time, Ms. Norelli-Bachelet is able to decode many of the enigmatic prophetic texts of the Indian Puranas as well as the biblical prophesies contained in the Revelation of Saint John the Divine. She has given a full account of John's heretofore impenetrable vision in her book, *The Hidden Manna*. The underlying themes of all apocalyptic prophesy is that a time will inevitably come when man is forced to step beyond the contradictions of the dualistic mental consciousness and into a consciousness of Unity. Before he does, however, he will come very near to destroying himself and the world. Ms. Norelli-Bachelet explains:

> *'The most intense struggle of the past 12,000 years has been experienced in our Manifestation, the 9th. The reason is obvious: the time of 'birth' is at hand. I have used the image of the Placenta to describe the crisis we are in the midst of because it allows us to appreciate, better than any other metaphor, the validity of the aphorism I have been employing over the past two decades: the Positive and the Negative serve the purposes of the One (see TNW, Volume 3). To really understand what I write one must know in every fibre of one's being that the Earth herself is a conscious, pulsating being. She possesses the same fourfold division as the individual, expressed geometrically in the Circle divided into Four: Physical, Vital, Mental, and Spiritual/Supramental. There is no difference at all between cosmic and individual. Then, if you can understand that the Child held within the Placenta is the One, nurtured in the cosmic Womb of the Mother over a period of 77,760 years to be precise, in applying the aphorism we realise that sustenance and nurturing have been accomplished by everything we have experienced on Earth over this period. What we call 'light' or 'dark' has no meaning other than as elements of hastening or restraining, all of which control the process so that Birth is neither premature and hence unviable, nor unduly retarded to result in a stillborn creation.'*[83] **Patrizia Norelli-Bachelet, *"Critical Threshold"***

> *'...We are now in the 9th Manifestation, equivalent to the ninth month when the child is born. On this position of the three-fold spiral, it is the time of the birth of that which was conceived some 54,000 years ago. But it must be pointed out that if man would know the past and future, he must be master of himself. If the time has come when the race as a*

> *whole is being given certain knowledge hitherto reserved for the enlightened, it is a sign that mankind on a collective scale has reached the point where it must become another more perfected species. If certain truths are handed to man it is because he must now become responsible and act in accordance with this greater knowledge.'* [84] **Patrizia Norelli-Bachelet, *The Gnostic Circle***

There can be no question that 'THE TIME HAS COME'. Every event, symbol, and symptom puts us in the throes of a Global transformation. Owing to the contradictions and paralysis of the reigning Mental consciousness, we are lost, bewildered and cannot find our way. Religion and Science continue to tighten their hold on a fearful humanity and, like a dog in a manger, do everything in their power to obstruct the movement, progress and dynamism required to realize this New World. But there is a new Prophesy and a new Myth that holds all of the secrets of the New Way, the new realization and the birth and development of a new world. It is the most exact and astonishing prophesy ever encapsulated in myth and follows the line of the Rig Vedic yoga. In her incomparable book, *The Magical Carousel*, Ms. Norelli-Bachelet provides us with a parable of the universal process by which individual souls take birth on Earth and grow into a more complete union with the Divine Consciousness upholding all Creation. Moreover, it allows individuals to find themselves within this Cosmic process, to understand the realities that they are dealing with and what the limitations are. It provides us with the context of UNITY in which there is no separation between the individual and the world and demonstrates how the progress of a single individual is linked with the whole human chain. The key feature of this New Myth and its Cosmological language is its Integrality, the sacred gift of the 3 in the Line. The Magical Carousel is Time in Motion, the highest expression of which is captured in the sacred dimensions of the Mother's Vedic Temple.

The Mother's Temple

What the Mother saw and recorded in those first days of 1970 was a masterpiece of correspondences, yet extremely simple in form. In the inner room we see the classic Vedic image of the Axis and the Plane that illustrates the alchemy of creation. The axis is seen as an immaterial solar ray of 15.20 meters that falls upon and into a 70 centimeter globe of light, the Center/Point/Womb rests upon a pedestal, a carved stone upholder inscribed with Sri Aurobindo's symbol. Beneath the pedestal is the Plane extending out horizontally in the image of the Vedic wheel which is also the Mother's own symbol. It is a play of perfect proportional relationships. The 24 hour day is contained in the 24 meters of the floor's diameter. The twelve months are represented by the walls and the 12 pillars that separate the inner core from the rest of the room. The year is written into the 15.20 Solar Ray/Shaft which strikes the core. The radius of the Sun, Earth and Moon are contained in its measurements as well as the movements of the great Precessional cycle that measures the Earth's history in terms of its spiritual destiny as it makes its orderly passage through the archetypal ages.

Inner Chamber and Core of the Mother's Temple

Patrizia Norelli-Bachelet
The New Way, Volume I & II,
Thea, Æon Books, 1981

Those with eyes to see will recognize the Mother's temple as the Temple of the Age. It is to our time what the Great Pyramid of Giza was to the Egyptians. Her plan was a miracle of sacred architecture in which everything is symbolic and everything a unity. The keys to Supramental Time are written in the inner room, and the secret formula (9-6-3-0/1) is given in its sublime measurements. Within these divine geometries are the lines of a new science, a spiritual physics that will open the way to a renaissance in understanding. It is, said the Mother, a 'Symbol of the Future Realization'.

The descent of Her Temple into the Earth plane may be seen as a re-creation of the 'churning myth' with Vishnu's line reestablishing the axis in creation to put a new order in the world. And if we recall Sri Aurobindo's wonderful description of our occult physiology, the temple is also an image of the same invisible Axis, Plane and Center within each individual, the self-same Soul discovered by the Rishis in the core of their own being. Mircea Eliade had told us earlier that if we could thoroughly understand just one level of correspondence that we would have a key to master the knowledge of all levels. But he would never have suspected that the one level that provided the key to all levels would be the human Soul. Sri Aurobindo describes some of the qualities of that luminous inner chamber:

> *'The body's rules bound not the spirit's powers:*
> *When life had stopped its beats, death broke not in;*
> *He dared to live when breath and thought were still.*
> *Thus could he step into that magic place*
> *Which few can even glimpse with hurried glance*
> *Lifted for a moment from mind's laboured works*
> *And the poverty of Nature's earthly sight.*
> *All that the Gods have learned is there self-known.*
> *There in a hidden chamber closed and mute*
> *Are kept the record graphs of the cosmic scribe,*

And there the tables of the sacred Law,
There is the Book of Being's index page,
The text and glossary of the Vedic truth
Are there; the rhythms and metres of the stars
Significant of the movements of our fate:
The symbol powers of number and of form,
And the secret code of the history of the world
And Nature's correspondence with the soul
Are written in the mystic heart of life. '[85]
***Savitri*, Book 1, Canto 5, CE, page 74-5.**

Eliade and the 'thrice great' Hermes would have been in awe of the Mother's Temple because its design spans the entire spectrum of correspondences from Macro to Micro. But, like all of the old spiritual realizers, they would only have been able to appreciate its static form revealing Being and its many correspondences. And this is where Sri Aurobindo, the Mother and Patrizia Norelli-Bachelet's work departs from the old static models of spirituality. The Mother's Temple is a temple of Time, of Movement. Its greatest mysteries lie not only in the Being, but in the Becoming of that Being which can only unfold in Time. Again, Sri Aurobindo explains:

'The significance of our existence here determines our destiny: that destiny is something that already exists in us as a necessity and a potentiality, the necessity of our being's secret and emergent reality, a truth of its potentialities that is being worked out; both, though not yet realised, are even now implied in what has been already manifested. If there is a Being that is becoming, a Reality of existence that is unrolling itself in Time, what that being, that reality secretly is is what we have to become, and so to become is our life's significance. '[86] **Sri Aurobindo, *The Life Divine*, Chapter 28 [underlining emphasis mine]**

In this passage Sri Aurobindo writes of something virtually unheard of in the old spiritual paths: Destiny, Meaning, Purpose, and that is precisely what is missing in both Science and contemporary Spirituality. If there is a destiny that is unrolling itself in time, a common destiny for both the world and the individual, that is what we need to know, to understand and align with. The New Way allows us to do this through an evolutionary model which is rooted in time and the Cosmic Harmonies, a path which respects the Being and is an unfailing instrument for its orderly Becoming.

During their time on Earth, Sri Aurobindo and the Mother worked tirelessly to perfect this knowledge and create appropriate forms for its physical manifestation. The most important of these was the Mother's plan for Her temple which was to be built according to the ancient science of *'Vaastu Vidya'* at the center of the microcosmic world which she and Sri Aurobindo had founded in Pondicherry. But owing to the un-centered blindness, resistance and ignorance of her disciples the Mother's plan for Her temple was systematically altered in favor of a version which glorified the stunted mental ego. One by one the elements of her Vedic design were distorted or discarded until not one was left intact. And when this abomination began to take form in concrete and steel, its mutilated dimensions began to produce calamitous effects in the field of its occult influence generating disunity and a palpable distortion of consciousness that continues on to this very day. The disciples, who rejected the Mother's original plan and those who continue to live under the toxic influence of the aberration built in her name, express an almost blind resistance to the new Supramental consciousness and will go to any extreme to suppress the knowledge of its existence. Because of the tamasic resistance of these so-called devotees who have usurped control of Sri Aurobindo's and the Mother's work and message, this impeccable Vedic truth remains unrecognized and unacknowledged and its incomparable influence on the Earth's evolution is virtually unknown.

If Sri Aurobindo and the Mother have given birth and architectural form to this new evolutionary consciousness, the unparalleled discoveries of Patrizia Norelli-Bachelet have given it a body, a vocabulary and a language, through which we may observe and understand its orderly deployment and its purpose in the world. Because of Ms. Norelli-Bachelet's efforts, it is now possible for the individual to not only have a direct and unequivocal experience of the Supreme Consciousness acting within the physical manifestation, but to SEE it as it is happening. Her gift of this New Seeing qualifies her as one of the world's greatest living treasures. But, like the Mother's Temple, she remains unrecognized and unacknowledged except for the few who are privileged to participate in her epochal work.

Conclusion

Among the most instinctive needs of the human being are order and meaning, a fixed point of reference that makes it possible to acquire some sense of orientation and vision in the chaotic world. Proverbs 29:18 tells us, '*...Where there is no vision, the people perish.*' We can see a poignant example of this truth in certain primitive cultures. In early tribal societies the elders and Shaman would erect a sacred pole/axis denoting the connection between the heavens and the Earth. The tribal community would live, dance and perform their rituals around this axis. If the lodge pole was broken or lost, the life of the community dissolved into chaos. Without that critical axis the tribe became aimless and disoriented. The people would give up all hope and prepare for death. Without a connection to a Divine order, life was simply not possible. One could say that the whole of human history revolves around this inherent need for order and orientation with the rise and fall of civilizations turning upon the nature of the axis and the quality of the vision it provided.

In the Vedic period we see a culture built around a similar center/axis but in the form of a temple whose sacred design incorporated the eternal principles of the cosmic order. The Rishis understood that if a culture were organized around a true center there would be harmony throughout the entire community. The Vedic temple text known as the *Mayamata* confirms the reality of this occult law of correspondence:

> *'If the measurement of the Temple is in every way perfect, there will be perfection in the universe as well.'*[87] ***Mayamata* XXII.92**

By following these ancient wisdom principles the Vedic forefathers enjoyed an extraordinary degree of harmony and integration in their culture. Following the design of the Vedic wheel, their society was organized on the basis of a unifying center and a fourfold arrangement of socio-economic categories called the 'Varnas'. According to Ms. Norelli-Bachelet, '*...The Vedic civilisation was founded on this system because it too arose from the same cosmic vision. In its fourfold order it mirrored the fourfold order of the universe.*'[88] The Varnas follow the same design as Twashtri's bowl, organizing the community as a whole into the four organic divisions of physical, vital, mental and spiritual. This ancient division of the Cosmic 'Purusha' or Person is known as the *'caste system'*.

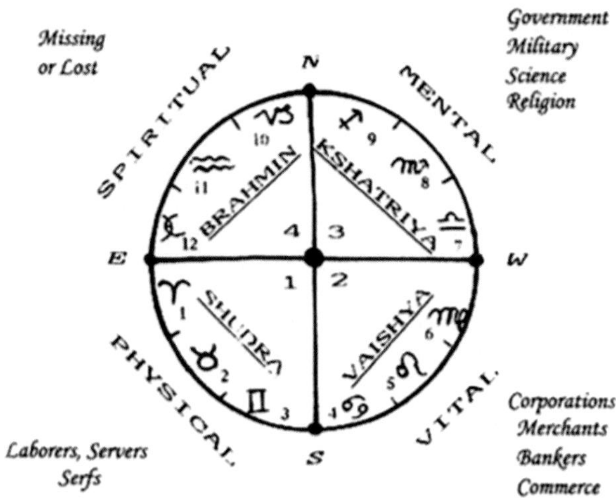

The Four Cosmic Divisions - *Varnas*, or Caste System

Caste	Zodiac
1) Shudra	1/Aries- 2/Taurus- 3/Gemini
2) Vaishya	4/Cancer- 5/Leo-6/Virgo
3) Kshatriya	7/Libra-8/Scorpio-9/Sagittarius
4) Brahmin	10/Capricorn-11/Aquarius-12/Pisces

In the Vedic period the 4 divisions were not arranged in a hierarchical order but in an integral circle because the Rishis knew that society, like the individual, could not function properly without the full contribution of each of its 4 parts. Under this arrangement everyone was valued and respected and had the same opportunity to perfect their particular destiny to the highest degree. But over the ages the original and cosmic sense of 'caste' suffered a tragic degeneration resulting in the rigid and dysfunctional social hierarchy that we see in the world today. But its basic divisions are still valid and when applied to our modern civilization they reveal precisely the same imbalances that we see in the individual; the Spiritual is missing or dormant and its indispensable role has been usurped by lesser functions. With a loss of the knowledge of the Seer, the

Mental/ego was seen as the highest and the circular organic division of society was replaced by a mental hierarchical order that remains in place today.

In the natural division of the four castes the *Brahmin* or Sage occupies the position of the highest spiritual wisdom. Members of the *Kshatriya* caste, which include the rulers, government and military officials, turn toward the Brahmin for advice and counsel because it is they who understand the deeper metaphysical complexities behind the apparent political realities. In the picture below we can see the Mother meeting with the three Prime Ministers of India for this purpose during their visit to the Sri Aurobindo Ashram on 29 September, 1955. From the left: the Mother, Jawaharlal Nehru (1947-64), Indira Gandhi (1966-77 & 1980-84), Lal Bahadur Shastri 1964-66) along with Kamaraj Nadar, President of the Congress (1967-71).

The Mother meeting with the three Prime Ministers of India during their visit to the Ashram in 1955

In the second picture India's former Prime Minister, Indira Gandhi is shown meeting privately with the Mother. Mrs. Gandhi was also in touch with Ms. Norelli-Bachelet for the same purpose that had to do with India's spiritual destiny and the role of her family in early nation-building.

The Mother and Indira Gandhi - 1971

With the loss of the wisdom of the Brahmin caste, rulers now turn to the *Vaishya* caste, which includes corporations, merchants and bankers in its ranks. In place of the Sage and Seer advising government, we now see 'Think Tanks' and pollsters and Wall Street executives whose underlying motives are not the good of the state but the insatiable pursuit of power and profits. Under this aberration which would never have been accepted by the Brahmin, we have seen a pillaging of the nation by special interests and corporate elites whose only interest in the working class is the means of their exploitation. The system is clearly broken and rushing headlong toward a disastrous reckoning.

Our global society has become like the tribe whose lodge pole was lost. There is no Center, no organizing Axis, simply the overwhelming chaos that Yeats described in his iconic poem. Instead of having a spiritual axis 'that holds', our great cities are built around commercial, technological or financial centers. And there is an unrelenting effort to organize the entire planet in that way. The two major world religions nearing the sunset of their existence are locked in a competitive, self-fulfilling, end-time theology in which each of their messianic saviors will only return following an Armageddon like battle of horrific proportions. And with the proliferation of nuclear technology, we grow nearer to that possibility every day.

If we take an honest and unvarnished look at what is going on around us we will know for a certainty that we are on an unsustainable course. The image that Ms. Norelli-Bachelet has used of a crushing birth process exactly describes our position. We are locked in a contracting womb. There is no Movement, no Progress, simply the building pressures of a disharmonious and disoriented humanity. Things left as they are can only worsen as the churning intensifies but the Mother offers us a choice:

> *'Before dying, falsehood rises in full swing. Still people understand only the lesson of Catastrophe. Will it have to come before they open their eyes to the truth? I ask an effort from all so that it has not to be. It is only the truth that can save us; truth in words, truth in action, truth in will, truth in feelings. It is a choice between serving the truth or being destroyed.'* [89]***A message issued by the Mother on 26 November, 1972***

For those whose look is clear the solution is obvious. If the principles of the Vedic culture were valid and true and created the means to establish harmony throughout their world, we need to embrace and restore those principles to our own age. The Mother clearly understood this need and gave out a plan that could unite the world and stand as the central axis of a new Supramental humanity. It was the Earth's Soul in the form of a Temple of the Age. The fact that it could not be built according to the stringent principles of the *Mayamata* shows just how misaligned and blind her disciples were, and are to this day. We simply cannot permit the false temple that was built in Her name to be passed off to the world as the Mother's own truth. It must be exposed for the fraud that it is.

> *'...The higher things, the guiding light we seek for humanity in the decades to come at this critical turning point in history has to be born of* ***the true Centre****. It cannot issue from a misaligned construction that pretends to occupy the centre on the basis of misrepresentation.'* [90] **Patrizia Norelli-Bachelet**

In order to remedy this regretful situation the sublime truths that the Mother's original plan contained must be rediscovered and rebuilt within each one of us. It is an effort that demands an increased aspiration, a desire for clarity and precision, and above all Knowledge - Veda. Moreover, it must be built on that sacred knowledge of Oneness, the only poise upon which the individual can forge a true Center and discover the living truth of that which is depicted so perfectly in the inner chamber of the Mother's Temple.

The options are straight-forward and clear. We can choose a New Way for humanity, the lines of which have been discussed herein or, like the Mother's disciples, we can choose the way of the Ego as it is reflected in the misaligned construction of Auroville's 'Shadow Temple'.

> *'...All those of good will who aspire to a new way for humanity, free from the suffocating hold of the Ignorance, must gather together and bring to light these matters so that we can truly begin to build a new world on the basis of a new world order such as the Mother had foreseen and fashioned in her unique cosmic Temple of the Age...'* [91] **Thea – Patrizia Norelli-Bachelet**

For those intrepid few who hear the call of such a destiny, everything turns upon an 'act of choosing' and surrender is an intrinsic part of that act. It is simply a state of being that unveils the Divine presence and allows us to share the vision of its becoming.

> *'...There is no moralistic judgment involved, no loss of one's soul and the like. There is simply an act of surrender and acceptance of the new, on its own terms. We either agree to make this* conscious *transition, or we remain imprisoned in the old and play out the collapse of a dying world. The choice we are faced with draws up from the depths of the intrepid seeker the finest energies of the true warrior of the Divine, for this dawning world belongs to the Earth's heroes of her cause.'* [92] **Thea – Patrizia Norelli-Bachelet, *Tenth Day of Victory*, Preface, page 15.**

RW

***

Notes:

A.) CERN scientists 99% certain they have found the Higgs Boson or God Particle. On July 4th, 2012, Rolf-Dieter Heuer, the director general of CERN, the multinational research center headquartered in Geneva and home to the Large Hadron Collider said 'I think we have it' as he announced the discovery of a subatomic particle that looks like the Higgs boson. According to the Standard Model, the Higgs boson is the only manifestation of an invisible force field, a cosmic molasses that permeates space and imbues elementary particles with mass. Particles moving through the field gain mass and without the Higgs field, as it is known, or something like it, all elementary forms of matter would be mass-less and there would be neither atoms nor life. One implication of their theory was that this cosmic molasses, normally invisible, would produce its own quantum particle if hit hard enough with the right amount of energy.

While many scientists compare the importance of this discovery to the Moon landing it really solves nothing, reveals nothing. Even now they are talking about there being more particles, maybe more forces to discover and the need to hit the Higgs field with more force to discover them. Since they are searching in vain, time will soon reveal that there is always another level to reach, another particle or force that according to science will explain everything if they simply throw enough force and money at the problem. In the meantime, critics of CERN are calling their announcement an intellectually dishonest charade designed to perpetuate their exorbitant funding at the expense of all other science, and an increasingly impoverished society. According to them, the search for the Higgs boson is 'a meaningless curve fitting exercise perpetuated by the greatest waste of funding that science has ever seen'.

*The LHC, built 330 feet underground along the French-Swiss border, began operations in September 10, 2008 when proton beams were successfully circulated in the main ring of collider. Nine days later operations were halted due to an accident resulting from an electrical fault. It was restarted but to run at only half-power.

A.1) Fresh data from the LHC suggest the Higgs boson is unlikely to pave the way to a profound new understanding of nature

Ian Sample, science correspondent
guardian.co.uk, Wednesday 14 November 2012 13.00 EST

Scientists working at the Large Hadron Collider have found no evidence that the new particle discovered earlier this year is anything but the simplest – and most boring – variety of Higgs boson.

Staff at Cern, the particle physics lab near Geneva, celebrated in July after they found what looked like the elusive boson amid the debris of scores of high-energy collisions inside the huge machine.

At the time, preliminary results from the two main experiments, Atlas and CMS, hinted that the particle might be something more exciting than the singular beast originally described in equations nearly 50 years ago. A more exotic Higgs could pave the way to a profound new understanding of nature.

But fresh data released by both teams at a conference in Kyoto today show that – so far at least – there is nothing peculiar about the particle's behaviour. The results do not completely rule out a more exotic Higgs particle, though. Some versions would look so much like the so-called Standard Model Higgs boson they could take years to identify.

B.) Spontaneous Formation of a 'Cosmos' In August of 2005 the *Aeon Centre of Cosmology* in collaboration with *The Movement for the Restoration of Vedic Wisdom* held a six-day Conference at Skambha, Ms. Norelli-Bachelet's estate in the Palani mountains just outside Kodaikanal India. Announcements and brochures had been sent out months before to present and prospective students of the New Way living all over the world. As the time of the conference arrived, a group of 4 students travelling from the United States left for Mumbai, India expecting a brief layover in Paris. When we arrived in

Paris the airline informed us that due to the heavy monsoons in Mumbai, the airport had been flooded and our flight would be delayed. Each day we returned to Charles De Gaulle airport only to be told that our flight was still on hold. On the third day we were cleared to continue our journey and we arrived at Skambha three days later than the other 12 participants who had already arrived. Out of the hundred or so brochures that had been sent out, when we assembled for our conference, 16 students had made the difficult trek to Skambha. Our numbers included Ms. Norelli-Bachelet, the 12 who had already arrived and the 4 who had come three days late. The Mother had mentioned that the presence of a person who had gone through the Supramental transformation was sufficient to create a 'cosmos'. But all of us were astonished to realize that we had been 'arranged' by that very power to perfectly form the structure of that cosmos as it has been illustrated in the Vedic Wheel, which is also the Mother's own symbol. We were 1, 4 and 12 making a total of 17. And just as in the alternating gender of the elements and zodiacal signs, there were 8 men and 8 women. The 12 had arrived first and in keeping with the law of 3 which relates to the gunas and time, the 4 of us making up the four inner petals of the wheel arrived 3 days after the 12. And of course Thea was the center of the wheel. When you are confronted with the existence of a consciousness that could arrange the schedules of 16 people travelling internationally, delaying 4 with a random monsoon in Mumbai so as to create a living Vedic wheel, it leaves absolutely no doubt as to who and what is in control.

***

References:

1. Sri Aurobindo, *India's Rebirth*, Institute for Evolutionary Research 1994, p. 94.
2. Sri Aurobindo, *The Secret of the Veda*, Collected Works of Sri Aurobindo, Sri Aurobindo Ashram Press, India, 1914 - 1916, p. 2.
3. Ibid, p 64.
4. Patrizia Norelli-Bachelet *'Sri Aurobindo and the Condition of Vedic Wisdom in India'*, February, 2007.
5. Sri Aurobindo, *Essays Divine and Human,* Collected Works of Sri Aurobindo, Sri Aurobindo Ashram Press, India, 1971, p. 68.
6. Sri Aurobindo, *Yoga of Self-Perfection,* Chapter XIX, 'The Nature of Supermind', *Arya* – July, 1920.
7. The Mother, *Mother's Agenda,* April 7, 1961, Institut de Recherches Evolutives: Paris, France.
8. Vaclav Havel, *Independence Hall Speech*, Philadelphia, July 4, 1994. Physics & Society Newsletter, October 1995.
9. Sri Aurobindo, *India's Rebirth*, Institute for Evolutionary Research 1994.
10. Arthur I. Miller, *Albert Einstein's special theory of relativity: emergence* 1905, Springer 1998.
11. Sri Aurobindo, *Letters on Yoga, December 2, 1946*, Sri Aurobindo Ashram Press, India, 1946, p. 212.
12. Will Keepin, *David Bohm: a life of dialogue between science and spirit*. Noetic Sciences Review. 30 1994.
13. Ralph T. H. Griffith, *Nasadiya Sukta - Hymn of Creation,* Hymns of the Rig Veda, X, 129, 1889.
14. Sri Aurobindo, *Synthesis of Yoga*, CWSA Vol. 23 – 24, Sri Aurobindo Ashram Press, India, 1999, p. 281.
15. The Mother, *Questions and Answers*, 11 April 1956, Collected Works of The Mother, Sri Aurobindo Ashram Press, India, p. 108.

16. Sri Aurobindo, *The Upanishads*; *'The Great Aranyaka'*, 'Horse of the Worlds', Sri Aurobindo Ashram Press, India, 1971. p. 275.
17. Stephen Hawking, *A Brief History of Time*, Bantam; 10th anniversary edition September 1, 1998.
18. Werner Karl Heisenberg, 1901 –1976, *The Uncertainty Principle*, Heisenberg 1927, cited in Mott & Peierls, 1977, p. 243.
19. Patrizia Norelli-Bachelet, *The Gnostic Circle,* pp. 120-122, and *The New Way Volume 2,* Chapter 4, pp. 260-261, Æon Books, New York.
20. Peter Plichta, Ph.D., God's Secret Formula - *Deciphering the Riddle of the Universe and the Prime Number Code,* Rockport, MA: Element Books, 1997.
21. Ralph T. H. Griffith, *Hiranyagarbha Sukta*, Hymns of the Rig Veda, X, 121, 1889.
22. Matsya Purana is the sixteenth purana of the sacred books of the Hindus and contains a comprehensive description of Manu and the Matsya Avatar.
23. Patrizia Norelli-Bachelet, *The Miracle of Oneness, 2012,* posted to Puranic Cosmology Updated, http://www.puraniccosmologyupdated.blogspot.com/.
24. Patrizia Norelli-Bachelet, 'Supermind and the Language of Gnostic Symbols', *The Vishaal Newsletter*, Vol. 9/3, August 1994.
25. Patrizia Norelli-Bachelet, 'Culture and Cosmos', *The Vishaal Newsletter*, Vol. 8 /2, June 1993.
26. Ezra Pound, Canto 85, *The Cantos of Ezra Pound,* London: Faber & Faber, 1956.
27. Sri Aurobindo, *Letters on Yoga*, December 2, 1946, Sri Aurobindo Ashram Press, India, 1946, p. 210.
28. Sri Aurobindo, In an interview on 8 May 1926, First appeared in the Sri Aurobindo Circle Eighth Number, 1952, pp. 99-102.
29. Nicola Tesla, In an interview with the New York Times in 1935.

30. Lee Smolin, *'The Life of the Cosmos'*, Oxford University Press, 1997.
31. Hans Christian Anderson, *'The Emperor's New Clothes'* first published in 1837.
32. Ervin Laszlo, *The Creative Cosmos: A Unified Science of Matter, Life and Mind,* Floris Books, Edinburgh 1993.
33. Roger Penrose, Interview concerning 'The Flow of Time' 1999, Moving Dimensions Theory. http://physicsmathforums.com.
34. Lee Smolin, *'The Life of the Cosmos'*, Oxford University Press, 1997.
35. William Butler Yeats, *"The Second Coming,"* written in 1919 and published in 1921 in his collection of poems Michael Robartes and the Dancer.
36. The Mother, *Collected Works of The Mother*, First Edition, Sri Aurobindo Ashram Press, India, Volume 09, pp. 296-301.
37. The Mother - *Questions and Answers*, May 5, 1929, Collected Works of The Mother, Sri Aurobindo Ashram Press, India.
38. *The Emerald Tablet of Hermes & The Kybalion*: Two Classic Books on Hermetic Philosophy, Create Space June 5, 2008.
39. Mircea Eliade, *Yoga: Immortality and Freedom*. 1958. New York: Pantheon Books.
40. *Ezekiel's Vision*, The Holy Bible, King James Version, Book of Ezekiel, Chapter 1, Verse 7.
41. Ralph T. H. Griffith, Hymns of the Rig Veda, I, 154, *The Three Steps of Vishnu Trivikrama*, 1889.
42. The Holy Bible, King James Version, *The Revelation of Saint John the Divine, Chap. 4; 1-9.*
43. Patrizia Norelli-Bachelet, *The Hidden Manna: Being the Revelation called Apocalypse of John the Divin*e, Æon Books, California, 1976, p. 35.
44. Ralph T. H. Griffith, *The Axle of the Earth,* Hymns of the Rig Veda, I, 164, 1889.
45. Ralph T. H. Griffith, *Twelve Spokes, one Wheel*, Hymns of the Rig Veda, 1.154.48; 1889.

46. Ralph T. H. Griffith, *The Vedic Wheel*, The Atharva Veda 10. 8, 1889.
47. Patrizia Norelli-Bachelet, *The Axis and the Plane, 2012,* Private circulation.
48. *The Churning of the Great Milk Ocean at the Dawn of Time*, The story appears in the Bhagavata Purana, the Mahabharata and the Vishnu Purana.
49. Patrizia Norelli-Bachelet, *Letter to Ramachandra Rao*, Private Circulation.
50. Sri Aurobindo, *The Growth of the Flame*, Savitri, A Legend and a Symbol, Canto Two, Sri Aurobindo Ashram Press, India, p. 370.
51. Sri Aurobindo, "The Synthesis of Yoga", Part Two, *the Yoga Of Integral Knowledge*, Sri Aurobindo Ashram Press, India, p. 362.
52. Patrizia Norelli-Bachelet, *The New Way, Vol. 3, 1983,* Æon Books, New York.
53. Ralph T. H. Griffith, *The Months and the Years*, Hymns of the Rig Veda, 2.24.5, 1889.
54. Ralph T. H. Griffith, *The Hidden Wheel,* Hymns of the Rig Veda, 10.85.16, 1889.
55. Ralph T. H. Griffith, *Twashtri's Bowl*, Hymns of the Rig Veda, 1.20.6, 1889.
56. Sri Aurobindo, *The Secret of the Veda*, Collected Works of Sri Aurobindo, Sri Aurobindo Ashram Press, India, 1914 - 1916, p. 341.
57. Patrizia Norelli-Bachelet, *The New Way, Vol. 1 & 2, 1983,* Æon Books, New York, p. 401.
58. Patrizia Norelli-Bachelet, *The Gnostic Circle,* Samuel Weiser Inc., New York, 1975, *pp. 110.*
59. Patrizia Norelli-Bachelet, *The New Way, Vol. 3, 1983,* Æon Books, New York, 1981.
60. Patrizia Norelli-Bachelet, Private Correspondence regarding '*Secrets of the Earth*' 2009.

61. Patrizia Norelli-Bachelet, *The Hidden Manna: Being the Revelation called Apocalypse of John the Divin*e, Æon Books, California, 1976.
62. Sri Aurobindo, *Letters on Yoga*, December 2, 1946, Sri Aurobindo Ashram Press, India, 1946, p. 251.
63. Ralph T. H. Griffith, *Usha the Dawn*, Hymns of the Rig Veda, 1. 123, 1889.
64. Sri Aurobindo, *The Life Divine*, Sri Aurobindo Ashram Press, India, 1946, p. 277.
65. The Mother, *Mother's Agenda, Institut de Recherches Evolutives*: Paris, France.
66. The Mother, *Mother's Agenda*, 28 November, 1964 *Institut de Recherches Evolutives*: Paris, France.
67. Sri Aurobindo, A. B. Purani, Evening Talks with Sri Aurobindo, The Second Series, 15-8-1925.
68. Stephen Hawking, *The Grand Design*, Stephen Hawking and Leonard Mlodinow, Bantam Books 2010.
69. The Mother, *Mother's Agenda, Institut de Recherches Evolutives*: Paris, France.
70. The Mother, *Collected Works of The Mother*, First Edition, Sri Aurobindo Ashram Press, India, Volume 09, pp. 296-301.
71. Sri Aurobindo, *Synthesis of Yoga*, CWSA Vol. 23 - 24 Sri Aurobindo Ashram Press, India, 1999, p. 439.
72. Sri Aurobindo, *Letters on Yoga*, December 2, 1946, Sri Aurobindo Ashram Press, India, 1946, pp. 509-10.
73. Sri Aurobindo, *The Life Divine*, Chapter XVI, 'The Triple Status of Supermind', Sri Aurobindo Ashram Press, India, 1946, p. 1016.
74. The Mother, *1915 letter to Sri Aurobindo*. Collected Works of the Mother, First Edition, SriAurobindo Ashram Press, India.
75. Sri Aurobindo, *Letter to the Mother,* Collected Works of The Mother, First Edition, , Sri Aurobindo Ashram Press, India, *1915*
76. The Mother, *Mother's Agenda*, 1962, Institut de Recherches Evolutives: Paris, France.

77. Patrizia Norelli-Bachelet, *'The Vedic Altar'*, 2012, p. 22, to be published.
78. Alchemical Axiom of Maria Prophetissa, Third Century, *Axiom of Maria*, Wikipedia Encyclopedia.
79. Sri Aurobindo, *The Life Divine*, 'The Supreme Truth Consciousness', Sri Aurobindo Ashram Press, India, 1946
80. Katha Upanishad, fifth century BCE.
81. The Mother, *Mother's Agenda*, 18 April 1961, *Institut de Recherches Evolutives*: Paris, France.
82. Sri Aurobindo, *The Life Divine*, Sri Aurobindo Ashram Press, India, 1946. p. 380.
83. Patrizia Norelli-Bachelet, *'Critical Threshold'*, Private Correspondence.
84. Patrizia Norelli-Bachelet, *The Gnostic Circle,* Samuel Weiser Inc., New York, 1975.
85. Sri Aurobindo, *Savitri, A Legend and a Symbol*, Book 1, Canto Five, Sri Aurobindo Ashram Press, India, pp. 74-5.
86. Sri Aurobindo, *The Life Divine*, Chapter 28, Sri Aurobindo Ashram Press, India, 1946, p. 1016.
87. The Mayamata, XXII.92 *Traité Sanskrit d'architecture*, written in the Tamil region during the, 11th century.
88. Patrizia Norelli-Bachelet, *'Caste, Culture and Cosmos'*, World Affairs, New Delhi,Volume 6, Number 1, Jan-Mar 2002
89. A message issued by the Mother on 26 November, 1972 just before a major cyclone hit Pondicherry on 5.12.1972.
90. Patrizia Norelli-Bachelet, *The New Way, Vol. 1& 2,* Æon Books, USA, 1981.
91. Patrizia Norelli-Bachelet,*The New Way, Vol. 1& 2,* Æon Books, USA, 1981.
92. Patrizia Norelli-Bachelet, *Tenth Day of Victory*, Æon Books, New York, 2003, Preface, p. 15.

Illustrations:

Cover. The Inner Chamber: photographs by Kai Sievertsen of William Netter's model of The Mother's original temple plan.

Title page, & page 55, The Vedic Wheel as described in the Rig Veda (1.154.48) and Atharva Veda (10.8). It may also be recognized as the Mother's Symbol. Mother's Symbol: Collected Works of the Mother, volume 13 "Words of the Mother", p.65, from the year 1955: 'The central circle represents the Supreme Mother, the Mahashakti. The four central petals are the four aspects of the Mother - and the twelve petals, Her twelve attributes.'

Page 59, The Churning of the Sea Milk, painting, Punjab Hills, 19th Century. This is an image of the Hindu Creation myth. It may also be found in Bas Relief at the temple in Angkor Wat, Cambodia.

Page 72, The Two Wheels; the Wheel of Space, Patrizia Norelli-Bachelet, *The New Way*, 'Aditi the Mother of Light', Volume I & II, Æon Books, 1981.P. 399. The Wheel of Time, Patrizia Norelli-Bachelet, *Symbols and the Question of Unity*, 'Language of the Universe' Sri Aurobindo Ashram Press, 1974, p. 114.

Page 75, Elements of this graphic first appeared in the Vishaal Newsletter and were later modified for use in a power point presentation for the Convergence Conference in Kodaikanal, India in 2008. The material is drawn from the Gnostic Circle. Patrizia Norelli-Bachelet, *The Gnostic Circle,* Samuel Weiser Inc., New York, 1975.

Page 78, Sri Aurobindo, Henri Cartier-Bresson's photograph of Sri Aurobindo in his study, Sri Aurobindo Ashram, Pondicherry, India, 1950.

Page 90, The Triangle of 9, 6, 3. Patrizia Norelli-Bachelet, *The Gnostic Circle,* 'The Circle Divided into Three Parts' *The Gnostic Circle,* Samuel Weiser Inc., New York, 1975.p.13. Patrizia Norelli-Bachelet, *Chronicles of the Inner Chamber*, Matrimandir Action Committee, 2007, p. 135, http://www.matacom.com/ .

Page 95, The Mother, giving Darshan to the devotees from the balcony adjoining her room. SriAurobindo Ashram, Pondicherry, India.

Page 104, Thea (Patrizia Norelli-Bachelet), Photograph, Patricia Heidt, On the porch at Skambha near the horses, 2005, India.

Page 105, The Map of the Twelve Manifestations, Patrizia Norelli-Bachelet, *The Gnostic Circle,*Æon Books, 1975.P. 19.Patrizia Norelli-Bachelet, *Symbols and the Question of Unity*, Sri Aurobindo Ashram Press, 1974, p. 33.

Page 109, The Inner Chamber of the Mother's Temple, Kai Sieversten: Graphic Artwork and photographs, William Netter, The Mother's Chamber. Patrizia Norelli-Bachelet, The New Way, Volume I & II, Æon Books, New York, 1981

Page 115, Twashtri's Bowl as the Caste System, THE VISHAAL NEWSLETTER, Volume 6, Number 4, October 1991 as a part of the series *Culture and Cosmos.* It has been reproduced for Puranic Cosmology Updated: http://puraniccosmologyupdated.blogspot.com/

Page 116, The Mother meeting with the three Prime Ministers of India during their visit to the Sri Aurobindo Ashram on 29 September 1955. From the left is the Mother, Jawaharlal Nehru (1947-64), Indira Gandhi (1966-77 & 1980-84), Lal Bahadur Shastri 1964-66) along with Kamaraj Nadar, President of the Congress (1967-71). The Map of India with the Mother's Symbol may be seen in the background.

Page 117, The Mother and Indira Gandhi. In 1971 Indira Gandhi was in a political turmoil because of the split in the Congress party's organization. Her government had lost its majority and on important occasions in the Parliament she relied on the support of the opposition group. She had ordered interim elections but thought she would be lucky if she could muster 250 seats in the house. It was at that time friendly advice brought her the suggestion that if she sought The Mother's support, her political and legislative uncertainty would end. Indira heeded the advice and came to Pondicherry to meet The Mother. Her prayer was for 250 house seats. The Mother smiled broadly, nodded her head vigorously and granted the prayer made through her cabinet colleague Nandini. The electoral victory was a landslide win. An Indira wave swept across the nation giving her 356 seats and the coveted two-thirds majority required to amend the constitution. Source -Wikipedia.

Thea (Patrizia Norelli-Bachelet) Selected Works

Books

The Magical Carousel – a Zodiacal Odyssey (1973), ISBN: 0945747306

Commentaries on The Magical Carousel (1979), ISBN: 0945747306

Symbols and the Question of Unity (1974), ISBN: 0-945747-54-3

September Letters (1974), no ISBN

The Gnostic Circle (1975), ISBN: 0-87728-411-3

The Hidden Manna (1976), ISBN: 0-945747-99-3

The New Way, Volumes 1 & 2 (1981) ISBN: 0-945747-06-3

The New Way, Volume 3 (2005) ISBN: 0-945747-03-9

Time & Imperishability (1997), ISBN: 0-945747-90-X

The Tenth Day of Victory (2003), ISBN: 0945747-33-0

Kashmir and the Convergence of Time, Space and Destiny (2004), ISBN: 978-0-945747-00-0

Explorations into the Leonardo Mania (2006), ISBN: 978-0-945747-05-5

Secrets of the Earth – Questions and Answers on the Line of Ten Avatars of Vedic Tradition (2009), ISBN: 978-0-945747-10-9

Newsletters/Chronicles:

The Vishaal Newsletter (1985-1995)

Chronicles of the Inner Chamber (2003), ISBN: 978-0-945747-81-9

The Partition of India - its cause, its purpose (2009)

Articles

'Numbers that Conquer (The Nehru Line)', Aside, Dec., 1984.

'In Defense of the Ancient Culture', The Hindu, 7 Nov. 2000.

'Vedic Astrology Isn't Vedic', Indian Express, June 3, 2001.

'Caste, Culture and Cosmos', World Affairs – The Journal of International Issues, (New Delhi), Vol. 6, No.1, Jan.-Mar. 2002.

'Cosmology in the Rigveda – The Third Premise', The Hindu, 9 July, 2002.

'Kashmir and the Convergence of Time, Space and Destiny, Prakash Bhagavaan Gopinath, Part I, April/June 2002, Part II, July/Sept 2002, Part III, Oct/Dec 2003, Part IV, Apr./June 2003.

'Explorations into the Leonardo Mania', New Dawn, Part I, Vol 101, Mar./Apr. 2007; Part II, Vol. 101, May/June 2007.

'India's True History is in its Myths', Bhavan's Journal, Vol. 54, No. 1, Aug. 15, 2007, Part II, Vol. 54, no. 2, Aug. 31, 2007.

'The Spirit That Moves Mountains', Naad - All India Kashmiri Samaj (AIKS), Vol 16, No.8, Aug. 2007.

'Hindu Restoration – Reconnect to Ancient Vedic Roots', Bhavan's Journal, Part I, Vol. 54, No. 10, Dec. 31, 2007); Part II, Vol. 54, No. 11, Jan. 15, 2008; Part III, Vol. 54, No. 12, Jan. 31, 2008; Part IV, Vol. 54, no. 13, Feb. 15, 2008.

'India and America – Part 1: Polar Opposites with a Common Role,' Bhavan's Journal, Vol. 54, No. 21, June 15, 2008; Part II, Vol. 54, No. 22, June 30, 2008.

'The Churning of the Nuclear Ocean', Bhavan's Journal, Part I, Vol. 55, No. 3, Sept. 15, 2008; Part II, Vol. 55, No. 4, Sept. 30, 2008.

'A Calendar That Unifies', Bhavan's Journal, Vol. 55, No. 18, Apr. 30, 2009.

Websites:

http://www.aeongroup.com/

http://www.rigvedacos.blogspot.com/

http://puraniccosmologyupdated.blogspot.com/

http://www.quantumyoga.org/index.html

The Convergence Conference 2008, Kodaikanal, India

About the author:

Robert E. Wilkinson– b. 1943, is an astrologer-cosmologist, and writer residing near Asheville, N.C. He has spent most of his adult life as a student of the Supramental Yoga established by Sri Aurobindo, the Mother and Patrizia Norelli-Bachelet, known to her students throughout the world as 'Thea'.
Mr. Wilkinson's primary area of interest is the role of applied cosmology in the areas of science, psychology and human development. Over the last 20 years Robert has worked closely with Thea and her students exploring the deepest aspects of the Supramental Yoga including a new model of Time as an aid in the realization of a Gnostic consciousness. Robert is one of the founding members of *Æon Group*, a non-profit educational organization dedicated to the promotion of the new Supramental Cosmology. He is also a founding member of the *Movement for the Restoration of Vedic Wisdom*, an international organization committed to the revival of the ancient Vedic knowledge and the reform of the Vedic/Hindu Calendar. An avid writer and blogger, Robert has published numerous articles on the subjects of Indian spirituality, yoga, sacred architecture, Vedic cosmology, and comparative mysticism.

For additional books and articles by Robert E. Wilkinson see: http://www.quantumyoga.org/RobertWilkinson.html

62838191R00077

Made in the USA
Columbia, SC
05 July 2019